Environmental Regimes in Asian Subregions

Environmental Regimes in Asian Subregions

China and the Third Pole

Simon Marsden

Professor and Chair in Energy Law, School of Law, Faculty of Arts and Humanities, University of Stirling, UK

Edward Elgar
PUBLISHING

Cheltenham, UK • Northampton, MA, USA

Published by
Edward Elgar Publishing Limited
The Lypiatts
15 Lansdown Road
Cheltenham
Glos GL50 2JA
UK

Edward Elgar Publishing, Inc.
William Pratt House
9 Dewey Court
Northampton
Massachusetts 01060
USA

A catalogue record for this book
is available from the British Library

Library of Congress Control Number: 2016953946

This book is available electronically in the **Elgar**online
Law subject collection
DOI 10.4337/9781784718091

FSC
www.fsc.org
MIX
Paper from
responsible sources
FSC® C013604

ISBN 978 1 78471 808 4 (cased)
ISBN 978 1 78471 809 1 (eBook)

Typeset by Columns Design XML Ltd, Reading
Printed and bound by CPI Group (UK) Ltd, Croydon, CR0 4YY

Contents

Preface and acknowledgements

This book had its origins in a previous Elgar project *Transboundary Environmental Governance in Asia: Practice and Prospects with the UNECE Agreements* (Marsden and Brandon, 2015). It builds on this earlier book, which focused on the UNECE *regional* environmental agreements with application in Asia as well as Europe, by evaluating *subregional* environmental agreements focused specifically on Asia. It develops this previous work by adding international relations and environmental management scholarship to international law, enabling interdisciplinary perspectives to inform the research and findings.

The current project considers nine existing regimes across southwest, central, southeast and northeast Asia, and proposes a tenth regime for the cross-subregion in south and east Asia. In addition to the focus on subregional agreements, two new issues are dealt with that were not addressed in the previous book: the role of China in relation to each of the agreements, and environmental governance of the Third Pole, the Himalayas/Tibetan Plateau. As an Asian subregion lacking in a specific environmental regime, the potential for this based on experiences in the other Asian subregions is analysed, and China's position is evaluated.

Preliminary research for the book was conducted in Hong Kong. I would like to thank Professor Gabriel Lau, Director, Institute of Environment, Energy and Sustainability, The Chinese University of Hong Kong (CUHK) for hosting my visit as a Senior Fellow from December 2014 until February 2015; and to Professor David Chen, then Director, Department of Geography and Resource Management (also CUHK) for arranging the visit, hosting a departmental seminar, and organizing library access in connection with this.

I would also like to thank two people in particular for their wisdom in relation to this project, Professor Falk Huettmann, University of Alaska, Fairbanks, who met with me twice after the Polar Law Symposium in 2015, and has provided input to my Third Pole interests; and notably, Professor Timo Koivurova, Director of the Arctic Centre, University of Lapland. Timo has been an unflinching supporter of all of my efforts to date and, despite the considerable distance between us and challenges of a very busy professional life, has always responded to requests for help at very short notice. Many thanks indeed to both of you.

Appreciation is also due to Professor Tim Stephens, ARC Future Fellow, Faculty of Law, University of Sydney, for kindly writing the Foreword at very short notice. I would furthermore like to thank the Elgar team (Ben Booth and Iram Satti in particular). Their support, encouragement, and a small extension of time to submit the final manuscript were invaluable to the completion of the book (and allowed me to move on to other tasks). I am finally especially grateful to the reviewers of the original and revised proposal for their consideration and wisdom, which improved the focus.

The manuscript was completed in July 2016 and is dedicated to my two youngest children, Harry and Isabel, from whom I was absent for too long during the work on the book. Thank you also Joseph and Anna – my other two children – for allowing me (mostly) the quiet time needed, and – as always – my partner in all my efforts in life, Liz.

Abbreviations

ACB	ASEAN Centre for Biodiversity
ACG	Azeri-Chirag-Guneshli
ADB	Asian Development Bank
AIIB	Asian Infrastructure Investment Bank
ALECSO	Arab League Educational, Cultural and Scientific Organization
AMME	ASEAN Ministerial Meeting on the Environment
ASEAN	Association of South East Asian Nations
BP	British Petroleum
CACO	Central Asian Cooperation Organization
CAMRE	Council of Arab Ministers Responsible for the Environment
CBD	Convention on Biological Diversity
CFE	Cooperation Framework on Environment
CIS	Commonwealth of Independent States
COBSEA	Coordinating Body on the Seas of East Asia
COP	Conference of the Parties
EA	Environmental Assessment
EAEC	Eurasian Economic Community
EANET	East Asia Network on Acid Deposition
EBRD	European Bank for Reconstruction and Development
ED	Executive Director
EIA	Environmental impact assessment
EMEP	European Monitoring [and] Evaluation Programme
EMMP	Environmental Mitigation and Management Plan
EU	European Union
GAINS-Asia	Greenhouse Gas – Air Pollution Interactions and Synergies

GCM	Global Climate Model
GEF	Global Environment Facility
GIS	Geographic Information System
GIWA	Global International Waters Assessment
GMS	Greater Mekong Subregion
GPA	Global Programme of Action
GTI	Greater Tumen Initiative
GTR	Greater Tumen Region
ICAS	Interstate Council on the Aral Sea
ICIMOD	International Centre for Integrated Mountain Development
ICJ	International Court of Justice
ICWC	Interstate Commission for Water Coordination
ICZM	Integrated Coastal Zone Management
IEE	Initial Environmental Evaluation
IFAS	International Fund for Saving the Aral Sea
IFI	International financial institution
IMO	International Maritime Organization
ISESCO	Islamic Educational, Scientific and Cultural Organization
IUCN	International Union for the Conservation of Nature
IWRM	Integrated Water Resources Management
IWT	Indus Waters Treaty
KAP	Kuwait Action Plan
LOSC	Law of the Sea Convention
MDB	Multilateral Development Bank
MDG	Millennium Development Goal
MEA	Multilateral environmental agreement
MEMAC	Marine Emergency Mutual Aid Centre
MOP	Meeting of the Parties
MOU	Memorandum of Understanding
MPA	Marine Protected Area
MRC	Mekong River Commission
NATO	North Atlantic Treaty Organization

NCP	Non-Compliance Procedure
NEASPEC	North-east Asian Subregional Programme for Environmental Cooperation
NGO	Non-governmental organization
NIR	National Implementation Report
NMC	National Mekong Committee
OPEC	Organization of the Petroleum Exporting Countries
PACS	Protected Areas of the Caspian Sea
PCA	Permanent Court of Arbitration
PERSGA	Programme for the Environment of the Red Sea and Gulf of Aden
PNPCA	Procedures for Notification, Prior Consultation and Agreement
RHAP	Regional Haze Action Plan
ROPME	Regional Organization for the Protection of the Marine Environment
SAARC	South Asian Association for Regional Cooperation
SACEP	South Asia Co-operative Environment Programme
SAS	South Asian Seas
SDG	Sustainable Development Goal
SEA	Strategic environmental assessment
SEPD	Subregional Environmental Policy Dialogue
TPE	Third Pole Environment
TRADP	Tumen River Area Development Programme
UNDP	United Nations Development Programme
UNDP-GEF	United Nations Development Programme – Global Environment Facility
UNECAFE	United Nations Economic Commission for Asia and the Far East
UNECE	United Nations Economic Commission for Europe
UNEP	United Nations Environment Programme
UNESCAP	United Nations Economic and Social Commission for the Asia Pacific
UNESCWA	United Nations Economic and Social Commission for Western Asia

UNWC	United Nations Water Convention
WB	World Bank
WSSD	World Summit on Sustainable Development

Foreword

This book puts paid to any suggestion that Asia is a barren field for the development of international environmental law. Asia is a dynamic and geopolitically pivotal region with rapidly expanding and changing economies and a rich variety of peoples, cultures and unique natural environments. Accompanying (though often lagging) the region's economic, social, cultural and political dynamism has been the emergence of a network of soft and hard law regimes for protecting the fragile and diverse ecosystems of the Asian subregions.

In this book Simon Marsden offers a deep and interdisciplinary analysis of the growing body of terrestrial and marine environmental regimes that span the tapestry of Asian political boundaries. The work is structured according to a systematic selection and analysis of environmental regimes in identified Asian subregions. Southwest Asia is the subject of Chapter 2, where the practical operation of the conventions and protocols in force for the protection of the Arabian Gulf and Gulf of Oman, and Red Sea and Gulf of Aden are examined. Chapter 3 shifts the focus to central Asia, concentrating on the arrangements in place for the inland seas of this subregion, particularly the Caspian and the Aral Seas (the latter of which has almost entirely disappeared). In Chapter 4 the attention moves eastwards to southeast Asia, which has been a laboratory for transboundary environmental governance, especially as a result of initiatives within ASEAN. Chapter 5 is devoted to an assessment of China's central role in transboundary environmental governance in the region as a whole, and in the 'Third Pole' in particular. Marsden explains the ecological and political significance of the Third Pole – the Tibetan Plateau that makes up a key component of the Earth's cryosphere and which spans south-western China (including the Tibet Autonomous Region), Nepal and parts of north-east India. This original discussion is a signal strength of this work, providing for the first time a comprehensive assessment of the legal and policy framework for the protection of an under-examined subregion. Marsden argues for the elaboration of a dedicated regime for the Third Pole that can address the many serious environmental impacts that are being felt here, particularly from massive dam-building programmes.

Environmental Regimes in Asian Subregions: China and the Third Pole provides a comprehensive and penetrating account of subregional international environmental law in Asia. It will undoubtedly become the standard reference for all those interested in the practical operation of transboundary environmental law in one of the world's most fascinating, complex and important regions.

Tim Stephens
Professor of International Law and ARC Future Fellow
Faculty of Law, University of Sydney, Australia

Table of international cases, treaties and other international instruments

TABLE OF INTERNATIONAL CASES

TABLE OF INTERNATIONAL TREATIES AND OTHER INTERNATIONAL INSTRUMENTS

1. Introduction: Asian subregions, environmental regimes and regime effectiveness

1. OVERVIEW

This chapter introduces the book, setting out the scope and aims, and outlines the research questions and methodology. Nine environmental regimes, together with other arrangements in Asian subregions, are evaluated and compared in the other chapters, most of which are situated within an international law framework. While this is the primary discipline on which the analysis is based, the book also draws on notions of effectiveness as developed by the international relations scholarship.[1] A common framework for analysing each of the regimes in Chapters 2–4 (and also one regime in Chapter 5), is explained and justified, with reference also to the scholarship of international political economy and global politics/geopolitics more broadly. This will assist with the future elaboration of a proposed international law framework for the protection of the environment of the 'Third Pole' in Chapter 5;[2] this draws on both hard and soft law, in recognition of existing treaty-based and other

[1] See for example, Alice Park Lee and Shunji Matsuoka, 'Effectiveness of International Institutions: How to Build an Analytical Framework?', paper presented to the *Conference on the Human Dimensions of Global Environmental Change*, Amsterdam, 2–4 December 2009; H Breitmeier, OR Young and M Zurn, *Analysing International Environmental Regimes: From Case Study to Database* (MIT Press, Cambridge, 2006); Edward L Miles, Steiner Andresen, Elaine M Carlin, Jon Birger Skjaerseth, Arild Underdal and Jorgen Wetterstad, *Environmental Regime Effectiveness: Confronting Theory with Evidence* (MIT Press, Cambridge, 2002); Oran R Young (ed), *The Effectiveness of International Environmental Regimes* (MIT Press, Cambridge, 1999). See also N Kanie, S Andresen and PM Haas (eds), *Improving Global Environmental Governance: Best Practices for Architecture and Agency* (Routledge, Abingdon, 2014).

[2] Falk Huettmann, 'Introduction: Why Three Poles and Why Protect Them', in Falk Huettmann (ed), *Protection of the Three Poles* (Springer, Japan, 2012) 3–32.

mechanisms. The Third Pole is constituted mainly by the Himalayas/Tibetan Plateau; in light of territorial, security and resource issues, such a proposal will inevitably be contested and controversial. Nonetheless, recommendations and suggestions for this are made in an effort to add a legal discourse to acknowledged environmental and other concerns.

Chapter 1 also outlines and distinguishes between regions and subregions[3] with reference to Asia,[4] and discusses regimes generally and environmental regimes specifically from the international law, international relations and international development literatures.[5] There is a need to both define and contextualize regions and subregions generally,[6] and specifically regimes and subregimes in Asia. The chapter therefore considers the geography of Asia as a region and the environmental issues it faces, before examining the numerous international and regional institutions that operate there. It also considers the discourse and scholarship in connection with regions and subregions developed by these institutions and academic commentaries, with a focus on the development of institutions to respond to the needs identified. It examines the (primarily

---

[3] For early reviews, see Luk Van Langenhove, 'Theorising Regionhood' UNU-CRIS Working Papers, W-2003/1 (United Nations University, Brugge, 2003); Rodrigo Tavares, 'The State of the Art of Regionalism: The Past, Present and Future of a Discipline' UNU-CRIS Working Papers W-2004/10 (United Nations University, 2004) 1–29; and Nikki Slocum and Luk Van Langenhove, 'The Meaning of Regional Integration: Introducing Positioning Theory in Regional Integration Studies' UNU-CRIS e-Working Papers W-2003/5 (United Nations University, 2003) 1–49.

[4] Note the work of the Asian Development Bank (ADB) in taking forward this agenda: Asian Development Bank, *Emerging Asian Regionalism: A Partnership for Shared Prosperity* (ADB, Manila, 2008); and with respect to central Asia, Asian Development Bank, Environment Division, *Central Asia Environments in Transition* (Manila, 1998).

[5] In an Asian context, see for example, 'Regime Theory' in H Litta, *Regimes in Southeast Asia: An Analysis of Environmental Cooperation* (Verlag Springer, Wiesbaden, 2012) 45–74.

[6] For recent reviews, see: Louise Fawcett, 'The History and Concept of Regionalism' UNU-CRIS Working Papers W-2013/5 (United Nations University, 2013) 1–19; Emil Kirchner and Roberto Dominguez, 'Security Governance in a Comparative Regional Perspective' UNU-CRIS Working Papers W-2013/8 (United Nations University, Brugge, 2013) 1–28; and Lorenzo Fioramonti, 'The Evolution of Supranational Regionalism: From Top-down Regulatory Governance to Sustainability Regions?' UNU-CRIS Working Papers W-2014/2 (United Nations University, Brugge, 2014) 1–26. The latter paper concludes (p18) that climate change and energy needs are as important as the global financial crisis to explain the resurgence of regionalism in recent times.

international relations) scholarship in connection with regimes and subregimes, before the notion and practicality of regime effectiveness (again primarily as advanced by international relations scholars), is considered.

The effectiveness of regimes under regime theory is discussed in some detail to establish the evaluation framework to be applied to the regimes in the substantive Chapters 2–4 that follow (and also to the one regime in Chapter 5). An outline of each of the chapters that constitute the book ends this first chapter. The analysis framework is developed with reference to international best practice, following an examination of the approach taken to multilateral environmental agreements (MEAs), in particular from the international law literature. The application of the United Nations Environment Programme (UNEP) support for MEAs is a key part of this framework; together with the UN Regional Commissions,[7] UNEP's role in regionalism has been significant, as seen in the Regional Seas Programme for example, on which three of the regimes in particular are based (the Arabian Gulf/Gulf of Oman, Red Sea/Gulf of Aden, and the Caspian Sea).[8] Another key part of the analysis framework draws on the findings of a related book project concluded in 2015;[9] with a concentration on implementation and compliance, findings from this with respect to National Implementation Reports (NIRs) and international Non-Compliance Procedures (NCPs) will also be focused upon in the framework. Finally, in recognition of the importance of international political economy[10] as well as regime theory, the policies and practices of multilateral development banks (MDBs)/international financial institutions (IFIs) are also included in the analysis framework. These are

---

[7] <http://www.regionalcommissions.org/> accessed 10/3/15. See also text below.

[8] <http://www.unep.org/regionalseas/programmes/default.asp> accessed 29/2/16. See generally Elizabeth Maruma Mrema, 'Regional Seas Programme: The Role Played by UNEP in its Development and Governance' in David Attard, Malgosia Fitzmaurice, Norman Martinez and Riyaz Hamza (eds), *The IMLI Manual on International Maritime Law, Vol III, Marine Environmental Law and International Maritime Security Law* (Oxford, Oxford University Press, 2016); and in connection with regime effectiveness, see also: David Johnson, Angela Benn and Maria Adelaide Ferreira, *Measuring Success: Indicators for the Regional Seas Conventions and Action Plans* (United Nations Environment Programme, 2014).

[9] Simon Marsden and Elizabeth Brandon, *Transboundary Environmental Governance in Asia: Practice and Prospects with the UNECE Agreements* (Edward Elgar Publishing, Cheltenham, 2015).

[10] Peter Newell, 'The Political Economy of Global Environmental Governance' (2008) 34(3) *Review of International Studies* 507–529.

significant because environmental protection requirements from international law are expected to be implemented in domestic legal systems prior to lending by MDBs/IFIs regardless of whether a state is a member of the relevant agreement or not.[11] Lending institutions have also been involved in the regime building itself, which has relevance for the Third Pole, in particular as seen in Chapter 5. Finally, in addition to its role in connection with each of the other subregions, the position of China is examined further in the last chapter.

2. SCOPE AND AIM OF THE BOOK

The book is an interdisciplinary analysis of inland, coastal and marine environmental regimes in Asia at the subregional level,[12] which are overwhelmingly transboundary in nature.[13] Particular disciplines are international environmental law, international relations and environmental management. These include regimes for the two UNEP 'Regional Seas' of southwest Asia (the 'Middle East'); the inland seas of central Asia (one part of this programme also); and rivers and their basins in southeast and northeast Asia. While still relevant, much of the previous scholarship in this area is now considerably dated,[14] and current work focuses on a limited range of regimes and comparisons.[15] Both past and current work

---

[11] Marsden and Brandon, above n. 9,

[12] For further examples, see Robin Warner and Simon Marsden (eds) *Transboundary Environmental Governance: Inland, Coastal and Marine Perspectives* (Ashgate, Farnham, 2012).

[13] Louis L Kotzé, *Global Environmental Governance: Law and Regulation for the 21st Century* (Edward Elgar Publishing, Cheltenham, 2012); and Louis L Kotzé and Thilo Marauhn (eds), *Transboundary Governance of Biodiversity* (Brill Nijhoff, 2014).

[14] Kazu Kato and Wakana Takahashi, *Regional/Subregional Environmental Cooperation in Asia* (Institute for Global Environmental Strategies, 2001); and Nathan Badenoch, *Transboundary Environmental Governance: Principles and Practice in Mainland Southeast Asia* (World Resources Institute, 2002). Notable exceptions include: Jörg Balsiger and Aysun Uyar (eds), *Comparing Regional Environmental Governance in East Asia and Europe: Proceedings* (Research Institute for Humanity and Nature, Kyoto, 2013); and H Litta, *Regimes in Southeast Asia: An Analysis of Environmental Cooperation* (Verlag Springer, Wiesbaden, 2012).

[15] H Litta, 'Comparison and Reflection: A Conclusion' in H Litta, *Regimes in Southeast Asia: An Analysis of Environmental Cooperation* (Verlag Springer, Wiesbaden, 2012) 189–200; in this case, water issues in the Mekong and transboundary haze pollution.

primarily concerns international law, relations and politics, and focuses on environment and development issues. The book updates this work by analysing regimes and other arrangements across southwest, central, southeast, east and south Asia; most have a legal underpinning of an international agreement, the majority of which are in force. It takes this further by adding recent information and analysis in connection with the most important recent player in many of these regimes, China. Based on the findings of each of the chapters, in Chapter 5 (south and east Asia), the book identifies the absence of a comprehensive regime for the Himalayas/Tibetan Plateau, notably for the protection of the mountains and related waters of what is known as the Third Pole.[16] Based on the experience of the other subregions, it makes suggestions for the development of such a regime.

The role of China in the existing regimes is examined at the end of Chapters 2, 3 and 4, and, with regard to east Asia specifically, in Chapter 5, whether as Party or Dialogue Partner to an agreement, or as financier, and/or developer in relation to the subregion in which the regime operates. China has a role to play as the state with the responsibility for the Tibetan Plateau – and interests across many of the regimes – it is therefore of key significance in the elaboration of a potential regime for the Third Pole, which is therefore examined based upon existing experience.[17] Together with India – which also shares many borders with other states in this subregion and has a similar focus on rapid development – cooperative efforts are urgently needed to address the implications of failing to develop

---

[16] Agreements for transboundary rivers are the exception in south Asia; India for example has adopted a number of treaties with its neighbours including the Indus Waters Treaty between the Government of India and the Government of Pakistan (Karachi, 19 September 1960), 419 UNTS 126, in force 1 January 1961. See Philippe Cullet and Joyeeta Gupta, 'India: Evolution of Water Law and Policy' in Joseph W Dellapenna and Joyeeta Gupta (eds), *The Evolution of the Law and Politics of Water* (Springer, 2009) 166–167.

[17] Simon Marsden, 'Developing Approaches to Transboundary Environmental Impact Assessment in China: Cooperation through the Greater Tumen Initiative and in the Pearl River Delta Region' (2010) 9(2) *Chinese Journal of International Law* 393–414; Simon Marsden, 'Assessment of Transboundary Environmental Effects in the Pearl River Delta Region: Is There a Role for Strategic Environmental Assessment?' (2011) 31(6) *Environmental Impact Assessment Review* 593–601; and Simon Marsden, 'Cooperative Environmental Planning in China's Pearl River Delta Region', in C Andressen, AR Mubarak and X Wang (eds), *Sustainable Development in China* (Routledge, Abingdon, 2013) 152–163.

such a regime.[18] These include the likely significant negative transboundary environmental effects of massive dam-building programmes of both nations and others, and the potential harm caused by exploration and exploitation of energy resources, as and when they are discovered and extracted.[19] The reason for the focus on China rather than India in this book is due to the vigorous external relations policy pursued by China in recent years; as seen, for example, in the establishment of the Chinese-led Asian Infrastructure Investment Bank (AIIB), and the development of the China-Pakistan Economic Corridor from Kashgar in Xinjiang, China to Gwadar in Pakistan on the Arabian Sea, via Pakistan-administered Kashmir. Both of these initiatives have potential to significantly impact upon the environment of the Third Pole, and are analysed in Chapter 5.

In relation to the existing subregional environmental regimes, the first and second outlined in Chapter 2 (southwest Asia) are the *Arabian Gulf/Gulf of Oman Convention*,[20] and the *Red Sea and Gulf of Aden Convention*,[21] which are focused primarily on coastal and marine issues, the latter also concerning the pan-region of Africa and Asia. There are eight western Asian Parties to the first: Bahrain, Iran, Iraq, Kuwait, Oman, Qatar, Saudi Arabia and the United Arab Emirates. There are now four western Asian Parties to the second: Jordan, Palestine, Saudi Arabia and Yemen. There are also two African Parties to the second: Sudan and Somalia. Saudi Arabia is therefore a Party to both of these agreements.

In Chapter 3 (central Asia) the third regime considered is shaped by the *Caspian Sea Convention*,[22] which concerns the pan-region of

---

[18] For an overview of the role of both powers, see: Christopher Ogden, *China and India: Asia's Emergent Great Powers* (Polity, Cambridge, 2016) in press; Christopher Ogden, *New South Asian Security: Six Core Relations Underpinning Regional Security* (Orient Black Swan, 2016); M Sornarajah and Jiangyu Wang, *China, India and the International Economic Order* (Cambridge University Press, Cambridge, 2010); and in relation to the NGO advocacy role, see Neda A Zawahri and Oliver Hensengerth, 'Domestic Environmental Activists and the Governance of the Ganges and Mekong Rivers in India and China' (2012) 12 *International Environmental Agreements* 269–298.

[19] Simon Marsden, 'Environmental Assessment of Cross-Border Development: China and the Third Pole' (2016) 18(1) *Journal of Environmental Assessment Policy and Management* 1650009-1–1650009-21.

[20] Regional Convention for Cooperation on the Protection of the Marine Environment from Pollution (Kuwait, 24 April 1978), in force 1 July 1979.

[21] Regional Convention for the Conservation of the Red Sea and Gulf of Aden (Jeddah, 1982), ECOLEX B7 p. 982:13, in force 20 August 1985.

[22] Framework Convention for the Protection of the Marine Environment of the Caspian Sea (Tehran, 4 November 2003) 44 ILM 1 (2005), in force 12 August 2006.

southwest,[23] central and northern Asia. This applies to the 'Caspian Littoral States' of Azerbaijan, Iran, Kazakhstan, Turkmenistan and Russia; all of these are southwest or central Asian states or have territory that extends into northern – or indeed eastern – Asia (Russia). The fourth regime examined is constituted collectively by the water resource arrangements for the Aral Sea: the *Joint Management and Conservation of Interstate Water Resources Agreement*,[24] the *International Fund for Saving the Aral Sea Agreement*,[25] and the *Interstate Commission for Water Coordination Statute*.[26] The Aral Sea lies to the east of the Caspian Sea between Kazakhstan and Uzbekistan, which are Parties to these bilateral arrangements. The fifth regime outlined is the *Environmental Protection for Sustainable Development Convention*;[27] this is not in force but has been signed by each of the five central Asia states: Kazakhstan, Kyrgyzstan, Tajikistan, Turkmenistan and Uzbekistan. Kazakhstan and Turkmenistan are therefore Parties to the first of these agreements and signatories to the second.

In Chapter 4 (southeast Asia) the sixth of the regimes analysed is the *Mekong Agreement*,[28] which has been ratified by Cambodia, Laos, Thailand and Vietnam; China and Myanmar additionally became Dialogue Partners in 1996. The seventh is the Association of Southeast Asian Nations (ASEAN) *Conservation Agreement*,[29] and related *Centre for Biodiversity Agreement*;[30] and the eighth is the ASEAN *Haze Pollution*

---

[23] The political geography of the Middle East has connections both with Africa and Asia, and appears unsettled.

[24] Agreement on Cooperation in the Field of Joint Management and Conservation of Interstate Water Resources (Almaty, 18 February 1992), <http://www.caee.utexas.edu/prof/mckinney/papers/aral/agreements/ICWC-Feb18-1992.pdf> accessed 27/1/16.

[25] Agreement about the Status of the International Fund for Saving the Aral Sea (Ashgabat, 9 April 1999).

[26] Statute of the Interstate Commission for Water Coordination of Central Asia (Almaty, 18 September 2008).

[27] Framework Convention on Environmental Protection for Sustainable Development in Central Asia (Ashgabat, 22 November 2006) TRE-143806, not in force, <http://www.ecolex.org/server2.php/libcat/docs/TRE/Full/En/TRE-143806.pdf> accessed 20/11/14.

[28] Agreement on the Cooperation for the Sustainable Development of the Mekong River Basin (Chiang Rai, 5 April 1995), unreported, in force 5 April 1995.

[29] Association of South East Asian Nations, Agreement on the Conservation of Nature and Natural Resources (9 July 1985) 15 EPL 64, not in force.

[30] Agreement on the Establishment of the ASEAN Centre for Biodiversity (Capitals of each member state, 2005).

Agreement.[31] While the *Conservation Agreement* is not in force and the *Haze Pollution Agreement* is, both have been approved by all or most of the ASEAN nations, who have also recently approved the *Centre for Biodiversity Agreement*.

In Chapter 5 (south and east Asia), existing governance arrangements are outlined further to the South Asia Co-operative Environment Programme (SACEP), and the South Asian Association for Regional Cooperation (SAARC). SAARC was formed by Charter 1985 by the following states: Bangladesh, Bhutan, India, Maldives, Nepal, Pakistan and Sri Lanka; Afghanistan joined later.[32] SACEP has the same membership. The latter has produced three relevant agreements, although these do not amount to an environmental regime, mainly because the environment is not the primary focus or detail is missing. Both these institutions and the agreements produced are however outlined because of their relevance to the establishment of a regime for the Third Pole. The agreements are the *Environmental Cooperation Convention*,[33] the *Rapid Response to Natural Disasters Agreement*,[34] and the *Energy Cooperation (Electricity) Agreement*.[35] Other mechanisms of note are either bilateral (*Indus Waters Treaty*),[36] or without hard law underpinnings (the *Malé Air Pollution Declaration*),[37] and the South Asian Regional Seas Programme.[38]

---

31 Association of South East Asian Nations, Agreement on Transboundary Haze Pollution (Kuala Lumpur, 10 June 2002), unreported, in force 2003.

32 Charter of the South Asian Association for Regional Cooperation (Dhaka, 18 December 1985).

33 SAARC Convention on Cooperation on Environment (Thimphu, 28–29 April 2010), unreported, in force 23 October 2013.

34 SAARC Agreement on Rapid Response to Natural Disasters (Maldives, 10–11 November 2011), unreported, not in force at November 2014, five states had ratified.

35 SAARC Framework Agreement for Energy Cooperation (Electricity) (Kathmandu, 27 November 2014), unreported, not in force.

36 Indus Waters Treaty between the Government of India and the Government of Pakistan (Karachi, 19 September 1960), 419 UNTS 126, in force 1 January 1961.

37 Malé Declaration on Control and Prevention of Air Pollution and its Likely Transboundary Effects for South Asia, 1998. See: <http://www.rrcap.ait.asia/male/Pages/Agreement.aspx> accessed 28/1/16.

38 <http://www.unep.org/regionalseas/programmes/nonunep/southasian/default.asp> accessed 8/6/2016.

The ninth regime is made up of the *Tumen Agreements*,[39] which were originally signed by North Korea, China and Russia, and later by South Korea and Mongolia. As with the South Asian Regional Seas Programme, the East Asian Regional Seas Programme is also not considered to constitute a regime as such because it lacks the treaty underpinning found in the western and central Asian arrangements. It is therefore only briefly outlined, together with the air pollution measures for northeast Asia. Similarly, the transboundary environmental governance for the jurisdictions of the Pearl River Delta 'Region' in southern China are noted but not discussed in detail.[40]

Chapter 5 finally focuses on the Third Pole as an additional tenth regime connecting south Asia (mainly the Himalayas) with east Asia (the Tibetan Plateau). Furthermore, there are connections with central Asia given the western extent of this high-altitude terrain. Based on the application of the developed analytical framework, the objective of the book is therefore to provide a comprehensive, comparative evaluation of the existing subregional environmental agreements in Asia, with a view to drawing lessons for other Asian subregions (or, indeed, globally), which do not have such a regime. The focus on the Third Pole as an additional regime emphasizes the links between environment, water and climate governance in particular, all of which have major transboundary dimensions. Based on the analysis of the existing regimes, the book therefore proposes a preliminary governance framework for the Third Pole, with a view to advancing the limited legal and policy discourse to date, and to gaining recognition and support for more detailed elaboration in the future.

3. RESEARCH QUESTIONS AND METHODOLOGY

The research approach is primarily a documentary analysis, informed where possible by informal, semi-structured interviews with experienced

---

[39] Agreement on the Establishment of the Tumen River Area Development Coordination Committee (New York, 6 December 1995), unreported, in force 6 December 1995; and Agreement on the Establishment of the Consultative Commission for the Development of the Tumen River Economic Development Area and Northeast Asia (New York, 6 December 1995), unreported, in force 6 December 1995.

[40] See Marsden, above n. 17, 2010, 2011 and 2013.

scholars. Documents evaluated include treaty texts and associated guidance, and materials produced by meetings of the Parties (MOPs); consultancy reports delivered in response to information needs identified by existing regimes; other international materials of a comparative nature, including multilateral and regional environmental regimes; and secondary source materials from scholars and non-governmental organizations (NGOs) closely connected with the regimes under review. From a review of the information obtained, including many case studies of the subregions' evaluation, the analytical framework was developed in ordered to assist with answering the research questions. Key questions were decided upon based on the findings of a review of the literature of environmental regimes and subregions. Other than the final two – transferability and the role of China in relation to the Third Pole, which will be answered in the concluding section of Chapter 5 – each of the others (including the role of China in connection with the other subregions), are answered in the concluding sections to Chapters 2–4. The research questions are as follows:

- What institutional arrangements are in place?
- What objectives and principles guide action?
- What are the key provisions?
    - o Are these institutional arrangements, objectives, principles and provisions effective?
- How can effectiveness be measured?
    - o In particular, what evidence of practice is there?
- Are more specialized arrangements present?
    - o In particular, to what extent are reporting mechanisms and compliance controls included?
- Are the institutional arrangements, objectives, principles and provisions transferable, or capable of transfer?
- What is China's role presently in relation to the environmental regimes?
    - o What can it potentially be in the future, particularly in relation to the Third Pole?

4. FRAMEWORK FOR ANALYSIS

As indicated in the overview above, the analysis framework is based on international best practice, with reference to the approach of MEAs and as reviewed by international environmental law scholars in particular.[41] From the literature it is clear that basic components must include the environmental duties found in general international law (no harm, cooperation, equitable utilization), environmental principles (precaution, polluter pays, etc.), substantive obligations, and treaty mechanisms to avoid and resolve disputes. However, compliance and enforcement, the acknowledged weakest elements in international environmental law, have been given particular emphasis in recent environmental treaties, as noted by the UNEP guidance in relation to compliance and enforcement. UNEP's Division of Environmental Law and Conventions has furthermore focused on strengthening links and synergies between MEAs as a means of promoting a systematic approach to coordination, and its guidance on compliance and enforcement remains very relevant today.[42] The need to include all such elements in any environmental regime, whether global or regional in application, is hence essential.

With respect to compliance and enforcement in the Asian context, however, it has been very clearly found that in many instances there has been a reluctance to accept the need for binding agreements at all, or to interfere in the environmental practices of other states.[43] This is in part as

---

[41] In addition to those outlined above n. 1, see also Adam Byrne, 'The 1979 Convention on Long-Range Transboundary Air Pollution: Assessing its Effectiveness as a Multilateral Environmental Regime after 35 Years' (2015) 4(1) *Transnational Environmental Law* 37–67, 39–40, distinguishing between legal (what is complied with), institutional (a cooperative approach) and normative (contribution towards justice, fairness and participation) effectiveness; Timo Koivurova, Pamela Lesser, Sonia Bickford, Paula Kankaanpää and Marina Nenasheva, *Environmental Impact Assessment in the Arctic* (Edward Elgar Publishing, Cheltenham, 2016) 15, regarding 'best practice research' focused upon what is implemented, and/or has been shown to work; and Timo Koivurova, Paula Kankaanpää and Adam Stępień, 'Innovative Environmental Protection: Lessons from the Arctic' (2015) 27 *Journal of Environmental Law* 285–311, 287–288 with respect to the need for organizational learning and adaptive governance.

[42] See Carl Bruch and Elizabeth Mrema, *Manual on Compliance with and Enforcement of Multilateral Environmental Agreements* (UNEP, Nairobi, 2006), available at: <http://www.unep.org/delc/MEAImplementationSupport/tabid/54401/Default.aspx> accessed 10/3/15. UNEP is not however without critics, see: DC Esty, 'Breaking the Environmental Law Logjam: The International Dimension' (2008) 17(1) *N.Y.U. Environmental Law Journal* 836, 846–848.

[43] Lorraine Elliot, 'East Asia and Sub-Regional Diversity: Initiatives, Institutions and Identity', in L Elliott and S Breslin (eds), *Comparative Environmental Regionalism* (Routledge, Abingdon, 2011) 60.

a result of the difference in Asian values,[44] or what has become known as the 'Asian', or 'ASEAN way'.[45] Indeed the importance of context is a readily acknowledged constraint upon any legal or other transplant.[46] The question of whether common standards and indicators are able to be applied to these different contexts has therefore been an issue for some time, whether used to evaluate environmental performance specifically, or notions of what amounts to good regional governance in general;[47] this is acknowledged by Best in the development of a set of parameters to evaluate the quality of governance in systems of regional cooperation or integration.[48]

Regime effectiveness is considered in more detail in a section below, and, as will be seen in the chapters that follow, the need for contextual perspectives remains to a considerable extent today. Indeed, the growing strength of a resource-hungry China in the region has been a cause for significant concern.[49] Dispute settlement in international fora for example is therefore frequently replaced by more cooperative approaches,[50] particularly where capacity may be lacking. As an alternative, the provision of assistance in order to avoid disputes is also likely to gain support and

---

[44] Roda Mushkat, *International Environmental Law and Asian Values: Legal Norms and Cultural Influences* (UBC Press, Vancouver, 2004)

[45] KL Koh and NA Robinson, 'Strengthening Sustainable Development in Regional Inter-Governmental Governance: Lessons from the "ASEAN Way"' (2002) 6 *Singapore Journal of International and Comparative Law* 640–682; see also Mark Beeson, 'The Coming of Environmental Authoritarianism' (2010) 19(2) *Environmental Politics* 276–294.

[46] Koivurova et al., 2016, above n. 41. In relation to EIA, context and effectiveness, see also Thomas Fischer (ed), *Environmental Assessment: Critical Concepts in Built Environment* (Routledge, Abingdon, 2015) volume 4, *Environmental Assessment Effectiveness*, Chapters 54–72, including Chapter 55, originally published as Simon Marsden, 'Importance of Context in Measuring the Effectiveness of Strategic Environmental Assessment' (1998) 16(4) *Impact Assessment and Project Appraisal* 255–266.

[47] Edward Best, 'Regional Integration and (Good) Regional Governance: Are Common Standards and Indicators Possible?' UNU-CRIS Working Papers, O-2004/6 (United Nations University, 2004) 1–45.

[48] Edward Best, 'The Assessment of Regional Governance: Principles, Indicators and Potential Pitfalls' UNU-CRIS Working Papers, W-2008/10 (United Nations University, 2008) 1–27.

[49] Simon Webb, 'ASEAN Says Seriously Concerned About Rising South China Sea Tensions', *Reuters*, 27 February 2016.

[50] Simon Chesterman, 'The International Court of Justice in Asia: Interpreting the *Temple of Preah Vihear* Case' (2015) 5(1) *Asian Journal of International Law* 1–6.

ensure that the objectives and outcomes of environmental agreements are fulfilled. The more modern approach of many of the UN Economic Commission for Europe (UNECE) agreements for example, in incorporating provision for NIRs and NCPs, has been found to be a positive support for environmental compliance in Asia.[51]

The question of effectiveness in relation to environmental regimes has been explored at length in the international relations and, to a lesser extent, international political economy literatures, and is also an important criterion for any analysis framework. However, previous subregional comparative analyses have not accorded significant weight to this, in recognition of the fact that 'It would be difficult to evaluate the effectiveness of these cooperative programmes/mechanisms against any reasonably acceptable measure of effectiveness'.[52] As an alternative, Kato therefore considered measures of achievement towards goals, objectives and targets, including action plans and projects;[53] this approach will also be employed in this book. The use of indicators has also been deployed recently in an Asian context, for example by ASEAN, where they have been used in the context of global administrative law.[54] It was found that the consensus approach to decision making prevailed, and that:

> Any attempt to incorporate the measurement of civil or political standards that lean towards a particular political ideology, into the ASEANstats indicators, would have met with strong disagreement and opposition. Further, such an attempt could be seen as a form of interference in the internal affairs of other states: for example, the measurement of press freedom might have been viewed by the more authoritarian states within ASEAN as advocating a more liberal model of governance. It is not unheard of for such political differences and contestation to suppress the expansion of an indicator, and this may thus

---

[51] Marsden and Brandon, above n. 9. With respect to water, see also Simon Marsden, 'The Helsinki Water Convention: Implementation and Compliance in Asia' (2015) 20(2) *Nordic Environmental Law Journal* 119–129; and Attila Tanzi, 'Regional Contributions to International Water Cooperation: The UNECE Contribution', in Laurence Boisson de Chazournes, Christina Leb and Mara Tignino (eds), *International Law and Freshwater: The Multiple Challenges* (Edward Elgar, Cheltenham, 2013).

[52] Kazu Kato, 'An Analytical Framework for a Comparative Study of Subregional Environmental Programmes in Asia', in Kato and Takahashi, above n. 14, 5.

[53] Ibid.

[54] Matthew Koh, 'Indicators in the ASEANstats Statistical Regime: A Case-Study on the Need for Accountability, Participation, and Transparency in International Governance by Indicators' (2016) 6(1) *Asian Journal of International Law* 159–191.

provide a reason for the unexplained, lopsided focus of the ASEANstats indicators.[55]

The remainder of this chapter will consider environmental regimes and subregimes, their effectiveness, and the Asian context for their operation, before ending with detailed chapter summaries.

5. REGIMES AND SUBREGIMES

A regime is a form of government or governance and combines law, politics and institutions. In more modern formulations, stakeholders may be both government and non-government organizations in the states concerned. A common dictionary definition for a regime is a 'set of rules, cultural or social norms that regulate the operation of a government or institution and its interactions with society'.[56] Although many dictionary definitions emphasize an authoritarian aspect, this is not necessarily seen in practice. However in the Asian environmental governance context, it has been recently argued that environmental authoritarianism may be better placed to deal with environmental issues than democratic governance, largely because of the pressures of population and development.[57] The word regime is also commonly used in international relations, politics and law to refer to more 'specialized arrangements', and often involves questions of sovereignty or jurisdiction.

Young comments: 'International regimes ... are more specialized arrangements that pertain to well-defined activities, resources, or geographical areas and often involve only some subset of the members of international society'.[58] These specialized arrangements for institutional components for example, may include MOPs, secretariat and working group functions, together with compliance procedures. Opportunities for other stakeholders to be involved, such as NGOs, individuals and the public, are also an important part of modern environmental regimes, in recognition of the fact that conventional interpretations of environmental governance based on regime theory are evolving, and 'critiques of mainstream approaches may yet help to articulate a pluralist account of global governance in which non-state actors play a more

---

[55] Ibid, 177, footnote removed from quotation.

[56] <http://en.m.wikipedia.org/wiki/Regime> accessed 10/3/15.

[57] Beeson, above n. 45, 276–294.

[58] Oran R Young, *International Cooperation: Building Regimes for Natural Resources and the Environment* (Cornell University Press, Ithaca NY, 1989) 13.

important part'.[59] Razzaque also comments in relation to the World Summit on Sustainable Development (WSSD): 'One of the major outcomes of the WSSD in 2002 was the emphasis on regional implementation though partnership between governments and civil society'.[60] And Söderbaum and Granit continue: 'There is also an emphasis on the role of stakeholders other than simply governments and regional organisations, which is considered important for recognising the multiplicity of relationships among countries, many of which involve the private sector and civil society'.[61]

Environmental regimes have been studied by international relations scholars for the last 25 years or more.[62] Research has distinguished between comprehensive environmental regimes which typically include all aspects of environmental control, and specific sub-environmental regimes which focus on a particular aspect, such as biodiversity conservation.[63] Kotzé comments: 'Regimes are among the institutional constructs that society has devised to respond to global environmental problems. They play a fundamentally important role in global (environmental) governance'.[64] In an international law context, framework agreements often provide for a comprehensive regime, which is then elaborated in more detail in protocols relevant to specific areas. As indicated above, regulation of the Caspian Sea is of this type, with a reasonably comprehensive treaty then detailed in a number of protocols which include both biodiversity conservation and environmental impact assessment (EIA).

Another example is specific governance for biodiversity conservation, which many be applied at either the multilateral, regional, or subregional levels. Within a state, national, regional and local efforts may in turn mirror global governance frameworks in order to implement international obligations; regional entities like the EU may also be required to

---

[59] Newell, above n. 10, 509.

[60] Jona Razzaque, *Environmental Governance in Europe and Asia: A Comparative Study of Institutional and Legislative Frameworks* (Routledge, Abingdon, 2013) 35.

[61] F Söderbaum and J Granit, *The Political Economy of Regionalism: The Relevance for International Waters and the Global Environment Facility* (Global Environment Facility, Washington DC, 2014) 23.

[62] Litta, above n. 5, 49.

[63] Kotzé (2012), above n. 13, 133.

[64] Ibid, 126.

implement international obligations.[65] Within Asia, MEAs include the *Biodiversity Convention*[66] and the *Climate Change Convention.*[67] Regional measures include the issue-focused treaties of the UNECE, the only UN Regional Commission to have produced any environmental agreements,[68] which includes the Helsinki *Water Convention.*[69] Sub-regional efforts include the agreements analysed in this book for the protection of the Arabian Gulf and Gulf of Oman, the Red Sea and Gulf of Aden, the Caspian Sea, the Mekong River Basin, and, to a lesser degree the Tumen region. They also include the conservation and haze pollution measures taken by ASEAN, and other applications by SACEP and SAARC. While formal regimes are prioritized because they are underpinned by legal mechanisms that typically apply to more than two states, other emerging regimes are also considered. These include other UNEP Regional Seas arrangements (in particular for south and east Asia), and air pollution measures in south and northeast Asia. In all cases, the following conclusion is generally true:

> In sum then, the broader environmental regime and its sub-regimes provide the opportunity and means for collective global action and a concerted regulatory response to environmental issues through various state and non-state actors and by means of a multitude of normative arrangements including many laws.[70]

---

[65] See for example Simon Marsden, 'Invoking Direct Application and Effect of International Treaties by the European Court of Justice: Implications for International Environmental Law in the European Union' (2011) 60(3) *International and Comparative Law Quarterly* 737–757, reprinted in JE Viñuales and E Lees (eds) *Environmental and Energy Law*, part of a three-volume collection, *The International Library of Law and Environment* (Edward Elgar, Cheltenham, in press).

[66] Convention on Biological Diversity (Rio de Janeiro, 22 May 1992), 31 ILM 822, in force 29 December 1993.

[67] United Nations Framework Convention on Climate Change (Rio de Janeiro, 9 May 1992), 1771 UNTS 107, in force 21 March 1994.

[68] Marsden and Brandon, above n. 9.

[69] Attila Tanzi, Owen McIntyre, Alexandros Kolliopoulos, Alistair Rieu-Clarke and Remy Kinna (eds), *UNECE Convention on the Protection and Use of Transboundary Watercourses and International Lakes: Its Contribution to International Water Cooperation* (Brill Nijhoff, Leiden, 2015); Tanzi (2013), above n. 51; and in an Asian context, see Marsden above n. 51.

[70] Kotzé, above n. 13, 134.

6. REGIME EFFECTIVENESS

Litta traces the history of environmental regime analysis, which 'evolved as a response to mounting international concern about threats to environmental quality since the 1970s'.[71] Many studies have focused on the effectiveness of environmental regimes, pointing to common features[72] which, in relation to issue specific regimes, 'sometimes play a decisive role in ameliorating environmental problems'. Common features of effective regimes, for example, are that benign problems are easier to solve than malign ones; that well understood problems are easier to solve than those that are clouded by uncertainty; and that the greater the problem-solving capacity and favourable political context, the greater likelihood of success.[73] Underdal supports these findings, evaluating effectiveness in accordance with two general propositions: the character of the problem itself, which may be easier to solve if it is intellectually less complicated or politically benign; and the problem-solving capacity, which will be greater if powerful institutional tools are available or because greater skill or energy is present to address it.[74] Litta considers the question of effectiveness at length in recent scholarship, asking the pertinent sub-questions as follows:

> Is it compliance with the regime rules? Is it the level of implementation of regulations stipulated by the regime? Is it the attainment of goals stipulated in the regime rules? Is it the achievement of desired outcomes with the least costs? Is it the peaceful agreement between stakeholders with different preferences? Is it the achievement of values [such] as fairness, justice, stewardship and equal participation? Or is a regime only effective if the quality of the environment is improved?[75]

It is of course all of these things, and the issue is more one of focus, given the practicalities of addressing them all. It is specifically a matter of disciplinary focus; dealing with the last question in particular requires an empirical dimension which in the environmental context is more likely

---

[71] Litta, above n. 5, 48.

[72] See for example, Young (1999), above n. 1; Miles et al. (2002), above n. 1; Breitmeier et al. (2006), above n. 1; Kotzé (2012), above n. 13.

[73] Miles et al., above n. 1, 'Introduction'.

[74] Arild Underdal, 'One Question, Two Answers', in EL Milev, A Underdal, S Andresen, J Wettestad, JB Skjærseth and EM Carlin (eds), *Environmental Regime Effectiveness: Confronting Theory with Evidence* (MIT Press, Cambridge, 2002) 3–45.

[75] Litta above, n. 5, 50.

to be considered by scientists than international law or relations scholars. This book has a more typical normative focus, in line with the comment by Newell that: 'It is not a caricature to suggest that environmental governance for many scholars of IR [International Relations] continues to be reduced to the study of international environmental *law* and the *institutions* that produce it and enforce it …'.[76]

7. ASIA AS HOME TO MANY ENVIRONMENTAL REGIMES AND SUBREGIMES

Asian states are Parties to numerous MEAs, including those highlighted above, and collaborate via these global regimes in addressing transboundary and other environmental challenges. As seen also above, the interest in regions and subregions has grown hugely over recent decades; this is a global phenomena with regional organizations found across the continents, and with inspiration drawn from the EU in particular, leading to significant EU/Asian comparative research.[77] Organizations in Asia considered in this book include the work of specifically environment-focused UN bodies such as UNEP, MDBs/IFIs such as the ADB, and economic cooperation bodies such as ASEAN; the latter two, as emphasized, also focus heavily on environmental concerns, in part because there is an economic cost to this. Gochhayat, in a recent study with special reference to India, provides a comprehensive introduction to notions of regionalism and subregionalism. The author begins with a definition which distinguishes between two very different meanings of regionalism:

> At the international level, regionalism refers to transnational cooperation to achieve a common goal or resolve a shared problem or it refers to a group of countries, such as Western Europe, the Western Balkans, or Southeast Asia, that are linked by geography, history or economic features. Used in this sense, regionalism refers to attempts to reinforce the links between these countries …
>
> The second meaning of the term … regionalism refers to a process in which sub-state actors become increasingly powerful and independent of the state: power devolves from the central state to regional governments within it. In

[76] Newell, above n. 10, 508 [original emphases].

[77] See for example, Razzaque, n. 60; also Jörg Balsiger and Aysun Uyar (eds), *Comparing Regional Environmental Governance in East Asia and Europe: Proceedings* (Research Institute for Humanity and Nature, Kyoto, 2013); and Tavares, above n. 3, who considers the notion of 'inter-regionalism' or cross-regionalism (at 19–20).

other words, it refers to a territory that is located within, or sometimes across the borders of a nation state.[78]

The focus of this book is upon the first of these meanings, while recognizing the relationship between them that comes from greater levels of integration leading to calls for more autonomy within, or indeed, separation from such groupings. If the EU is the best example of the former meaning, it also admirably demonstrates the latter. Increased integration has produced calls from the UK to renegotiate terms of the treaties upon which the EU is founded, as well as calls to leave the EU; if this is to occur, there are however genuine concerns about the effect on the environment.[79] At a different level, calls for autonomy by regions and subregions within member states of the EU – such as Scotland within the UK,[80] and Catalonia within Spain[81] – illustrate their autonomous claims and how regionalism and subregionalism are fluid concepts, which may or may not be realized.[82]

The UN has emphasized the importance of regions and subregions in the work of its Statistics Division,[83] Environment Programme,[84] and via its Regional Commissions.[85] In 1999, the Statistics Division developed a system of macro-geographical (continental) regions, subregions and other selected economic groups to report advances towards achieving the Millennial Development Goals (MDGs) globally.[86] The UN Geoscheme

---

[78] A Gochhayat, 'Regionalism and Sub-Regionalism: A Theoretical Framework with Special Reference to India' (2014) 8(1) *African Journal of Political Science and International Relations* 10.

[79] Fiona Harvey, 'Brexit Would Damage UK Environment, Say Experts', *The Guardian*, 27 January 2016; for a contrary view, see George Monbiot, 'Why the EU's Increasing Failure to Protect Nature Means I May Vote No', *The Guardian*, 10 July 2015.

[80] Ruth Wishart, 'The EU Referendum is Already Following the Scottish Playbook, Project Fear 2.0', *The Guardian*, 26 February 2016; Severin Carrell and Libby Brooks, 'Nicola Sturgeon: Second Scottish Independence Poll Highly Likely', *The Guardian*, 25 June 2016.

[81] Giles Tremlett, 'Catalonia's Separatist Government will Struggle to Break Stalemate with Spain', *The Guardian*, 11 January 2016.

[82] Note the position with respect to Hong Kong; SC Yeung, 'Is HK Independence a Mainstream Thought Among the Youth?' *Economic Journal Insight*, 24 February 2016.

[83] <http://unstats.un.org/unsd/default.htm> accessed 10/3/15.

[84] <http://www.unep.org/> accessed 10/3/15.

[85] Above n 7.

[86] <http://millenniumindicators.un.org/unsd/methods/m49/m49regin.htm#asia> accessed 10/3/15. In connection with this, see Emanuela Maria Cavaleri,

distinguishes these regions and subregions into 'Central Asia', 'Eastern Asia', 'Southern Asia', 'South-Eastern Asia' and 'Western Asia', each identified with a numerical code and composed of states as indicated.[87] Some of the divisions and compositions may appear inaccurate geographically,[88] ethnically/culturally,[89] or are indeed politically controversial.[90]

In 2003, the Asia Pacific region of UNEP held its first Subregional Environmental Policy Dialogue (SEPD) meeting in Beijing; these have been held annually since. UNEP is the Secretariat for the annual SEPD, facilitating the preparation of background documentation and providing logistical support for/in liaison with the host government, as well as documentation of the meeting records and recommendations. The functions of UNEP are therefore to: provide a forum for consolidating Asia-Pacific views on global environment issues; deliver regional inputs to MEAs and global events such as UNEP's Governing Council and the Global Ministerial Environmental Forum; advise on critical and emerging environmental issues in Asia and the Pacific; and provide policy guidance for the effective implementation of programmes in Asia and the Pacific. There are five subregions of the UNEP Asia Pacific region: central Asia, south Asia, southeast Asia, northeast Asia and the south Pacific; western Asia is also a focus of UNEP efforts, with various institutional mechanisms established; finally, the Regional Seas Programme is often noted as one of its major achievements.

'The Post-2015 UN Development Framework: Perspectives for Regional Involvement' UNU-CRIS Working Papers, W-2014/12 (United Nations University, 2012) 1–28.

[87] <http://unstats.un.org/unsd/methods/m49/m49regin.htm#asia> accessed 10/3/15.

[88] Russia is not included as an Asian state, despite the majority of its territory (east of the Urals) constituting North Asia, for which there is no category; Turkey by contrast is included in Western Asia, despite (admittedly very small) parts of its territory being within Europe. Iran is also included within Southern Asia not Western Asia.

[89] Cyprus, Israel and Georgia are included in Western Asia despite their close involvement/association with Europe. Northern Cyprus, with the greatest cultural and ethical claims to being part of Western Asia, is a Turkish Republic not recognized by the international community.

[90] The Hong Kong and Macau Special Administrative Regions are both separately included in Eastern Asia but not Taiwan. Furthermore, the State of Palestine is now included in Western Asia alongside Israel; the existence of both is contested by various states.

There are five UN regional commissions, of which three are active in environmental policy making in Asia: the Economic and Social Commission for the Asia Pacific (UNESCAP), the Economic and Social Commission for Western Asia (UNESCWA), and the UNECE. The UNECE Environmental Policy is the oldest and has promoted sustainable development further to Agenda 21; a Committee was formed from senior advisors to ECE governments in 1971, with the terms of reference from the 62nd session in 2007 including a focus on the Caucasus and central Asia. In relation to UNESCAP, a 1st Ministerial Conference on Environment and Development was held in 1985, which has been held every five years since. Regarding UNESCWA, a Regional Coordination Mechanism for Arab States held its 20th Meeting in December 2014.

The UNECE is particularly relevant because it has a large number of Asian members;[91] as indicated, it is also the only one of the regional commissions to have produced any environmental treaties (5) or related protocols (12),[92] several of which have been inspirational for the subregional regimes in their own development. For example, the *EIA Convention*[93] has been a model for the *EIA Protocol*[94] under the *Caspian Sea Convention* and is assisting states Parties to the *Mekong* and *Tumen* regimes in elaborating more detailed measures; and the *Public Participation Convention*[95] is also playing a role in considering best practice requirements under the *Caspian Sea* and *Mekong* arrangements.[96]

---

[91] Turkey, Israel, Russia, Georgia, Armenia, Azerbaijan, Kazakhstan, Kyrgyzstan, Tajikistan, Turkmenistan and Uzbekistan.

[92] Wiek Schrage, Keith Bull and Albena Karadjova, 'Environmental Legal Instruments in the UNECE Region' (2008) 18(1) *Yearbook of International Environmental Law* 3–31; and Marsden and Brandon, above n. 9.

[93] Convention on Environmental Impact Assessment in a Transboundary Context (Espoo, 25 February 1991) 30 ILM (1991) 802, in force 27 June 1997.

[94] Protocol on Environmental Impact Assessment in a Transboundary Context to the Framework Convention for the Protection of the Marine Environment of the Caspian Sea, 2012; see also associated *EIA Guidelines*.

[95] Convention on Access to Information, Public Participation in Decision-Making and Access to Justice in Environmental Matters (Aarhus, 25 June 1998) 2161 UNTS 447, in force on 30 October 2001. This is often known as the 'Aarhus Convention'.

[96] With respect to the former, see Framework Convention for the Protection of the Marine Environment of the Caspian Sea, Conference of the Parties Fifth Meeting [29–30 May 2014, Ashgabat] Turkmenistan, Items 5 and 12 of the Provisional Agenda, Recommendations and Analysis Based on the Comments by the Aarhus Convention Secretariat on Public Participation under the Tehran Convention, Note by the Interim Secretariat, TC/COP5/INF.6.

The distinction between UN regions and subregions as drawn by the Statistics Division, UNEP and the Regional Commissions highlights the very high degree of regionalism and subregionalism within Asia. The ADB commented extensively upon this in a 2008 study, which was focused on economic development of Asia's regions and subregions. Not surprisingly, this gave minimal consideration to environmental issues, although energy demands and implications, and impacts upon water and air, were noted.[97] It commented: 'Asian regionalism is the product of economic interaction, not political planning'.[98] This recognized how, initially at least, all the drivers were economic, in more recent times to meet the challenges of the 1997/98 Asian financial crisis. The geographical scope of the report is furthermore limited to the 16 economies of the ASEAN states, India and others in east Asia, with west, central, north (Russia in particular) and south Asia (except India) excluded.

With a focus on the political economy of regionalism, other authors have considered environmental effects in more detail, for example the relevance for international waters.[99] Söderbaum and Granit comment for instance that '"Regionalism" refers to the *common objectives, values and identities* that lead to region-formation and regional cooperation within a given geographical area. It often leads to the creation and development of regional institutions and regional governance frameworks in order to shape and regulate collective action'.[100] Other approaches to regionalism include: by geography, for example the distinction between greater China (including Taiwan) and mainland China, what is considered part of northeast Asia, and also north Asia; by geology, for example the 'Third Pole' (Himalayas/Tibetan Plateau); by eco-regions, for example river basins, such as the Mekong; by politics, economics, or security, for example the role of ASEAN and the Eurasian Economic Community; by religion, consider the Middle East/western Asia; and, finally, by language and culture – consider Cantonese, spoken by people across southern China despite jurisdictional divides.[101]

---

97 Asian Development Bank, above n. 4, 207–210.

98 Ibid, 26.

99 Söderbaum and Granit, above n. 61.

100 Ibid, 7 [original emphasis].

101 Note that ironically 'Canton' is the generic name also for Swiss subregions.

8. DETAILED CHAPTER SUMMARIES

By way of an overview, following this Chapter 1, Chapters 2–5 will analyse the subregional agreements in Asia – grouped into southwest (Chapter 2), central (Chapter 3), southeast (Chapter 4), and south and east (Chapter 5) Asia. As will be seen in the latter chapter, the development of subregional agreements is much less developed in east and south Asia, and it is for these subregions that a proposed regime for the Third Pole is considered. Technically encompassing territory in east Asia (the Tibetan Plateau), as well as the mountains in the south Asian states, the Third Pole is a *cross-subregion* in the same way that the First Pole (the Arctic), is a *cross-region* (north America, northern Europe, north Asia).

The role of China in east Asia and more generally is also considered in Chapter 5, following analysis of its role in the other subregions outlined at the end of the preceding chapters. This role is either as a Party or Dialogue Partner, international lender or developer. Apart from the questions answered in Chapters 2–4, Chapter 5 also contains other conclusions in response to the research questions outlined above, recommendations and consideration of future research directions. For each of the regimes considered in Chapters 2–4, and where possible, also for the suggested regime in Chapter 5, objectives, principles, procedures and other mechanisms are evaluated with reference to the key research questions and international best practice; evidence of practice to demonstrate effectiveness will be indicated wherever possible. In more detail, chapter summaries are as follows.

Chapter 2 concerns southwest Asia. It analyses the *Arabian Gulf/Gulf of Oman Convention*, and the *Red Sea/Gulf of Aden Convention*, together with the numerous detailed protocols that has been elaborated thereunder. The objective of both is the conservation of the Arabian Gulf and Gulf of Oman, and Red Sea and Gulf of Aden environments respectively, including the prevention, abatement and combating of marine pollution. Each contains general and specific obligations, including the prevention of pollution from ships, from dumping, and from land-based sources; from exploration and exploitation; and from other human activities. They furthermore contain obligations to conduct EIAs and for liability and compensation.

Chapter 3 examines central Asia. It first evaluates the *Caspian Sea Convention*. The objective of this is to protect the Caspian environment from all sources of pollution including the protection, preservation, restoration and sustainable and rational use of the biological resources. It contains guiding principles and obligations such as EIAs. It also evaluates the water resource arrangements for the Aral Sea: the *Joint Management and Conservation of Interstate Water Resources Agreement*, the *International Fund for Saving the Aral Sea Agreement*, and the *Interstate Commission for Water Coordination Statute*. In addition, Chapter 3 considers the *Environmental Protection for Sustainable Development Convention*. The purpose of this last treaty is to ensure the effective environmental protection and improvement of the environment, including the rational use of natural resources, and to reduce and prevent transboundary environmental damage through harmonization and coordination. Principles and provisions for specific environmental media are included, together with cooperation, access to information and public participation.

Chapter 4 is focused on southeast Asia. It explains and analyses the *Mekong Agreement*. This is focused on cooperation in relation to sustainable development, and the utilization, management and conservation of the water and related resources to optimize the multiple-use and mutual benefits of all riparians, and to minimize harmful effects. Obligations on the Mekong River Commission (MRC) include creating a Basin Plan, and on Parties to notify and consult in relation to development, principally dam-building. Chapter 4 will also explain and analyse the ASEAN *Conservation Agreement* and related *Centre for Biodiversity Agreement*. The first contains many broad environmental objectives and obligations, as well as specific provisions on conserving genetic diversity, endangered and endemic species, vegetation cover and forest resources, soil, water and air. EIAs, advance notification, and appropriate consultation should all be undertaken prior to any proposed utilization of the shared resources. Chapter 4 finally evaluates the *Haze Pollution Agreement*. This aims to prevent, monitor and mitigate negative transboundary effects arising from land and forest fires through national efforts and international cooperation. Measures to control the sources of fires; develop monitoring, assessment and early warning systems; exchange information and technology; and provide mutual assistance, are all included.

Chapter 5 looks to south and east Asia. It considers the environmental governance currently in place in south Asia, in particular as supported by SACEP and SAARC. The latter has produced three relevant agreements: the *Environmental Cooperation Convention*, the *Rapid Response to Natural Disasters Agreement*, and the *Energy Cooperation (Electricity) Agreement*. These are all considered along with the *Indus Waters Treaty*, climate change arrangements, the *Malé Declaration* on air pollution, and the Regional Seas Programme for the South Asian Seas. Significantly, as indicated, it also evaluates the SAARC and SACEP institutional arrangements that operate independently of the agreements promulgated, which may form the basis for any Third Pole regime. In east Asia, it furthermore analyses the *Tumen Agreements*. The aim of both is to attain environmentally sound and sustainable development of northeast Asia and the Tumen River Economic Development Area, with reference to international and domestic law and the requirements of MDBs/IFIs. Obligations are also present for EIA and environmental management plans. Chapter 5 also examines the informal arrangements for air pollution in northeast Asia, and the UNEP Regional Seas Programme for East Asia.

Significantly, Chapter 5 analyses the prospects for establishing an additional environmental regime for the Third Pole, and China's role in relation to this. This is based on the conclusions from each of the preceding chapters and in the context of the particular environmental issues and the geopolitics of this subregion. While most of the regimes in the subregions discussed in Chapters 2–5 are based on formal agreements, and the majority have been adopted (and in many instances ratified and applied with some – albeit limited – success), the governance of the Third Pole is yet to be conceived, fully proposed and evaluated. Finally, Chapter 5 answers the remaining central research questions not answered in the previous chapters and summarizes findings. It also emphasizes the likely challenges. These include – despite the significance and necessity of engaging China (and other key states) in environmental regimes, and particularly in the Third Pole subregion – that a reluctance to cede national sovereignty to international law-making may impede progress. Nonetheless, other successful examples of Chinese involvement in the implementation of and compliance with international law (such as the *World Heritage Convention*),[102] together with shared interests between China and other states in resolving the major environmental issues of the twenty-first century (climate change uppermost), are indications of promise for the future. Research directions going forward are elaborated at the end of Chapter 5.

---

[102] Convention Concerning the Protection of the World Cultural and Natural Heritage (Paris, 16 November 1972) 1037 UNTS 151, in force 17 December 1975.

2. Southwest Asia: the Arabian Gulf/Gulf of Oman and the Red Sea/Gulf of Aden regimes

1. INTRODUCTION

In this chapter, the first two of the environmental regimes identified will be outlined and analysed, with a particular focus on implementation and compliance in accordance with the framework for analysis described in Chapter 1. The same analytical approach will be applied to each of the regimes in this chapter and in Chapters 3, 4 and 5; these are also focused on existing environmental regimes in the central and southeast Asian subregions, together with one regime in east Asia. This approach will enable comparisons to be drawn for the purpose of responding to the research questions in Chapter 1, which include lessons learnt for the elaboration of a prospective regime for the Third Pole in Chapter 5.

The chapter will proceed as follows. After concluding the introduction with a definition of southwest Asia, the first section will analyse the *Kuwait Regional Convention for Cooperation on the Protection of the Marine Environment from Pollution* (*Arabian Gulf/Gulf of Oman Convention*) and related protocols;[1] the second section will analyse the *Regional Convention for the Conservation of the Red Sea and Gulf of Aden Environment* (*Red Sea/Gulf of Aden Convention*) and related protocols.[2] In relation to the evaluation of each, subsections will provide geographical context, historical background and institutional arrangements; objectives, principles and an overview of procedures including matters of

[1] Kuwait Regional Convention for Cooperation on the Protection of the Marine Environment from Pollution (Kuwait, 24 April 1978) in force on 1 July 1979 (the 'Arabian Gulf Convention' is also known as the 'Kuwait Convention').

[2] Regional Convention for the Conservation of the Red Sea and Gulf of Aden (Jeddah, 14 February 1982), ECOLEX B7 p. 982:13, in force 20 August 1985, (the 'Red Sea/Gulf of Aden Convention' is also known as the 'Jeddah Convention').

implementation and compliance; and, where possible, evidence of practice will be incorporated. Comparisons between the *Arabian Gulf /Gulf of Oman* and the *Red Sea/Gulf of Aden* agreements (and related protocols) with be highlighted in these sections, before reference is made to China's role, with conclusions to follow at the end.

The southwest Asia regimes are considered in date order rather than in accordance with their geographical position in Asia (from west to east), because the *Arabian Gulf/Gulf of Oman Convention* and its protocols have been the inspiration for the *Red Sea/Gulf of Aden Convention* and its protocols. It is apparent from a reading of both, however, that the provisions of both treaties, and the protocols they have in common, are almost identical; hence the discussion of the *Red Sea/Gulf of Aden Convention* is much briefer than the *Arabian Gulf/Gulf of Oman Convention*, with a focus on identifying any differences of note.

1.1 Defining Southwest Asia

Southwest Asia has not been comprehensively defined by the UN, but is the preferred descriptor for this subregion because of its geographical precision. While there may not be a corresponding subregion known as 'northwest Asia',[3] to term the more southerly subregion simply 'western' Asia ignores a geographical reality; it is also inconsistent with the existence of a northeast and southeast Asia.[4] Nonetheless 'Western Asia' is the term used by the UN Geoscheme to define the territory encompassing 18 countries: Armenia, Azerbaijan, Bahrain, Cyprus, Georgia, Iraq, Israel, Jordan, Kuwait, Lebanon, Oman, Palestine, Qatar, Saudi Arabia,

[3] The reason for this appears to be because the boundary between the European and Asian continents in Russia is demarcated by the Ural Mountains which are to the north of Kazakhstan in central Asia; therefore eastern Europe (where it is part of Russia) takes the place of northwest Asia.

[4] Note however that the reason why Chapter 4 (southeast Asia) is so called is because it is a very precise, commonly used descriptor. In contrast, south and east Asia, which are dealt with together in Chapter 5 (since environmental regimes are limited in each and because this is where the Third Pole is found), are so called because of the blurred boundaries of both of these subregions. In relation to south Asia, it is extremely uncommon to include the southeast Asian states within south Asia; in relation to east Asia, the term enables consideration of a broader subregion than that usually referred to as northeast Asia, which in itself often excludes Asian Russia, the most easterly part of the entire Asian continent.

Syria, Turkey, the United Arab Emirates and Yemen.[5] These include Mediterranean (Cyprus, Turkey and Israel), Caucasian (Armenia, Azerbaijan and Georgia) as well as the Arabic countries. The total population is about 300 million, with Turkey and Iran (about 75 million each) being the most populous, followed by Iraq and Saudi Arabia (about 32 million each). The International Union for the Conservation of Nature (IUCN) limits the number of states to 13,[6] and comments that 'West Asia is part of three eco-regions that incorporate a variety of ecosystems, including Mediterranean forests, deserts, plains, rangelands, savannas, oases, mountains, rivers, lakes, springs, mudflats, swamps, marshes, mangroves, seagrass beds and coral reefs, all of which provide a wide array of ecosystem services and ecological corridors for migratory species through the region'.[7]

The Council of Arab Ministers Responsible for the Environment (CAMRE) is the subregion's principal environmental ministerial forum.[8] UNEP supports CAMRE to develop regional environmental programmes and promote regional mechanisms for sustainable development. Through CAMRE and other specialized Ministerial Councils, and in cooperation with international, regional and Arab organizations, the League of Arab States has adopted the Arab Initiative for Sustainable Development.[9] This identifies the challenges towards achieving sustainable development in the subregion, and gives priority to: combating land deterioration and desertification; integrated management of water resources; and integrated management of coastal zones and marine resources.[10] The Gulf Cooperation Council is also of relevance, and its Environmental Affairs Committee.[11] The task of this committee is to deal with all environmental

---

5 <http://unstats.un.org/unsd/methods/m49/m49regin.htm> accessed 14/5/15.

6 Adding Iran, which under the UN Geoscheme is part of South Asia, but excluding the Caucasian states of Armenia, Azerbaijan and Georgia, and also excluding Turkey, Israel and Cyprus in the eastern Mediterrean.

7 <https://www.iucn.org/regions/west-asia/about-rowa> accessed 14/11/16.

8 <http://www.arein.org/en/council-of-arab-ministers/> accessed 14/5/15.

9 <http://www.arein.org/en/objectives/> accessed 14/5/15.

10 United Nations Environment Programme, 'The Regional Perspective', *UNEP in 2003*, UNEP, 2003.

11 See generally and for collaboration with the World Bank <http://www.worldbank.org/en/country/gcc> accessed 14/3/16; and Stephen Lintner, Sherif Arif and Mario Hatziolos, *The Experience of the World Bank in the Legal, Institutional and Financial Aspects of Regional Environmental Programs: Potential Applications of Lessons Learned for the ROPME and PERSGA Programs* (World Bank, Washington DC, 1996). Note also the contribution of the UN Global Environment Facility in building cooperation and resolving conflict

issues of interest to the member states. It has adopted a unified code of standards for the protection of the environment in the subregion.

From a socio-political or economic perspective, it has been said that the majority of these states, which are part of the 'Middle East', are a 'region without regionalism', and have 'the lowest relative degree of regional integration in the contemporary world'.[12] Coskun points to the absence of common goals and organizational development as key indicators that are absent in the subregion, despite the commonalities of a national lack of democratic traditions and relatively weak civil society, as added to the – for the most part – independent post-colonial state experience of each following the two World Wars.[13] Despite development and linkages since, and notable uprisings collectively known as the 'Arab Spring',[14] it is likely that much of this remains true today. Nonetheless, what has also been termed the 'Greater Middle East' in recognition of historical connections with north Africa, Turkey, the Caucasus, central Asia, and Afghanistan and Pakistan in south Asia, is without doubt a politically strategic and economically significant region where religion, nationalism and conflict – among many other things – continues to influence global affairs.[15]

---

around international waters. See: Andrea K Gerlak, 'One Basin at a Time: The Global Environment Facility and Governance of Transboundary Waters' (2004) 4(4) *Global Environmental Politics* 108–141.

[12] Aarts and Kuhnhardt, cited by Bezen Balamir Coskun in 'Region and Region Building in the Middle East: Problems and Prospects' UNU-CRIS Working Papers, O-2006/1 (United Nations University, 2006) 1, 1.

[13] Ibid, 3.

[14] See for example: Adam Roberts, 'The Arab Spring: Why Did Things Go So Badly Wrong?' *The Guardian*, 15 January 2016, which highlights the internationalized civil wars in Syria and Yemen, and rise of Islamic State; and for an overview, <http://www.bbc.com/news/world-middle-east-12813859> accessed 04/03/16.

[15] M Parvizi Amineh (ed), *The Greater Middle East in Global Politics: Social Science Perspectives on the Changing Geography of the World Politics* (Brill, Leiden, 2007).

2. *ARABIAN GULF/GULF OF OMAN CONVENTION* AND ASSOCIATED PROTOCOLS

2.1 Geographical Context, Historical Background and Institutional Arrangements

The eight state Parties to the *Arabian Gulf/Gulf of Oman Convention* are: Bahrain, Iran, Iraq, Kuwait, Oman, Qatar, Saudi Arabia and the United Arab Emirates; all are in southwest Asia. The area regulated by the *Arabian Gulf/Gulf of Oman Convention* is endowed with valuable natural resources and significant biodiversity of plant and animal species.[16] These include wetlands, waterfowl, mangroves, fish, marine mammals, turtles, corals and other forms of life, with the dugong population second only in global importance to Australia's. There are also around twenty species of dolphin and whale, all five subtropical species of turtles, and more than a thousand species of fish, most endemic and with a high commercial value.[17] Sheppard et al. in a comprehensive review published in 2010 outline the main benthic habitats, ecosystems and species groups found that the Arabian Gulf;[18] they find, however, that the Gulf is changing rapidly, primarily as a result of significant oil- and gas-sponsored economic development. Despite all the evidence available at the time of writing, they comment:

> Changes are so rapid in this region, however, that many accounts are soon overtaken by developments which include construction, substantial coastline alterations, habitat loss, creation of beds of shifting or suspended sediments,

---

[16] Arabian Gulf/Gulf of Oman Convention, above n. 1, Article 2; <http://www.unep.org/regionalseas/programmes/nonunep/ropme/default.asp> accessed 14/11/16.

[17] See: Dalal Al-Abdulrazzak, Dirk Zeller, Dyhia Belhabib, Dawit Tesfamichael and Daniel Pauly, 'Total Marine Fisheries Catches in the Persian/Arabian Gulf from 1950 to 2010' (2015) 2 *Regional Studies in Marine Science* 28–34.

[18] Charles Sheppard, Mohsen Al-Husiani, F Al-Jamali, Faiza Al-Yamani, Rob Baldwin, James Bishop, Francesca Benzoni, Eric Dutrieux, Nicholas K Dulvy, Subba Rao V Durvasula, David A Jones, Ron Loughland, David Medio, M Nithyanandan, Graham M Pilling, Igor Polikarpov, Andrew RG Price, Sam Purkis, Bernhard Riegl, Maria Saburova, Kaveh Samimi Namin, Oliver Taylor, Simon Wilson and Khadija Zainal, 'The Gulf: A Young Sea in Decline' (2010) 60 *Marine Pollution Bulletin* 13, 16–22.

and temperature and salinity changes in restricted water flows along the coast, as well as by climate warming.[19]

The impacts of land-based activities on the coastal waters are significant. These include municipal sewage and industrial effluents from petroleum refineries and power, desalination and petrochemical plants; and dredging and reclamation activities. To emphasize significance, a Global Programme of Action (GPA) for the Protection of the Marine Environment from Land-Based Activities has been in place for some time.[20] Kirk notes the 2006 Report on the State of the Marine Environment produced for the Second Intergovernmental Review Meeting of the GPA, which found that progress in dealing with the nine GPA source categories has been uneven. Progress has been made in relation to persistent organic pollutants, radioactive substances and hydrocarbons. However, results were mixed for heavy metals and sediments; and had worsened in connection with sewage, nutrients, litter, and physical alteration and destruction of habitats.[21] Pressures on fisheries, which are an important contribution to food security, and on marine mammals are reported as a direct consequence of development and, indirectly, climate change, which has impacted these species and invertebrates, often through algal blooms, degraded water quality, and resulting marine mortality.[22]

Operational and accidental oil pollution is another major challenge in the subregion,[23] as are the impacts on the marine environment by offshore oil installations, whatever international obligations may require

---

19 Ibid, 15.

20 Elizabeth A Kirk, 'Noncompliance and the Development of Regimes Addressing Marine Pollution from Land-Based Activities' (2008) 39(3) *Ocean Development and International Law* 235, 239.

21 Ibid.

22 Sheppard et al., above n. 18, 22–25.

23 See generally: Don Hinrichsen, *Coastal Waters of the World: Trends, Threats and Strategies* (Island Press, Washington DC, 1998), Chapter 14, 'The Arabian Gulf: Politics and Oil'; and specifically in relation to offshore oil pollution, Nengye Liu, 'Protection of the Marine Environment from Offshore Oil and Gas Activities' in Rosemary Rayfuse (ed), *Research Handbook on International Marine Environmental Law* (Edward Elgar, Cheltenham, 2015); land and vessel source pollution and ocean dumping are also covered by separate chapters in this edited book. Similarly, another recent edited volume contains similar chapters – see David Attard, Malgosia Fitzmaurice, Norman Martinez and Riyaz Hamza (eds), *The IMLI Manual on International Maritime Law. Vol III, Marine Environmental Law and International Maritime Security Law* (Oxford University Press, Oxford, 2016), in particular Chapters 3–5.

of owners and operators.[24] The dumping of ballast water and pollution from ships, are therefore a major cause of chronic oil pollution.[25] In relation to the former, the International Maritime Organization (IMO) entered an agreement with the UN Development Programme – Global Environment Facility (UNDP-GEF) to establish the Global Ballast (GloBallast) Water Management Project, which was approved in 1999 and was aimed at controlling a ship's ballast waters and sediments. Nadim et al. comment: 'By and large, there is a need for a cooperative, multi-lateral, regional approach in the Persian Gulf region. No one port or even one Gulf country can address this problem alone'.[26] In relation to the latter, the Mumbai High North disaster on 27 July 2005 in the Arabian Sea, around 160km west of the Mumbai coast, which killed 22 people, is cited as one of the ten worst disasters of its kind.[27] In relation to collisions at sea,[28] Nadim et al. give an example from the Arabian Gulf on outcomes and implications from incidents of this kind:

> On September 6, 2005, a nuclear submarine collided with a Turkish cargo ship. Although minimal, these kinds of accidents often occur in the waters off the ports in the Gulf region and point to the fact that naval accidents can have catastrophic outcomes. If a naval accident led to leaking of radioactive or hazardous materials into the Gulf, the desalination plants in the area would have to shut down their operations indefinitely. In the case of accidents occurring in the Strait of Hormuz, there could be a disruption in the flow of oil from the Persian Gulf[,] sending the price of oil to levels that could swiftly create a worldwide economical crisis.[29]

---

[24] Hossein Esmaeili, *The Legal Regime of Offshore Oil Rigs in International Law* (Ashgate, Dartmouth, 2001).

[25] <http://www.unep.org/regionalseas/programmes/nonunep/ropme/> accessed 19/3/15.

[26] Farhad Nadim, Amvrossios C Bagtzoglou and Jamshid Iranmahboob, 'Coastal Management in the Persian Gulf Region within the Framework of the ROPME Programme of Action' (2008) 51 *Ocean and Coastal Management* 556, 560.

[27] Offshore technology.com, 'The World's Worst Offshore Oil Rig Disasters' (2 January 2014) <http://www.offshore-technology.com/features/feature-the-worlds-deadliest-offshore-oil-rig-disasters-4149812/> accessed 04/3/16.

[28] The definitive text on this largely private law area is Simon Gault, Steven Hazelwood, Andrew Tettenborn, Stephen Girvin, Stewart Buckingham, Thomas Macey-Dare and Edward Cole (eds), *Marsden and Gault on Collisions at Sea* (Sweet and Maxwell, London, 14th edition, 2016).

[29] Above n. 26, 559.

Unfortunately, incidents continue to occur, and issues with offshore installations are also highly significant. As seen elsewhere in the world, such as the Gulf of Mexico[30] and the Timor Sea,[31] these offshore blowouts are often the result of regulatory failings. However, offshore operations involve considerable risk; extracting petroleum requires oil-based drilling muds and chemicals, producing harmful emissions. Vinogradov explains:

> There are four main stages of offshore petroleum development: geological and geophysical surveying; exploration; development and production; and decommissioning. Each of these stages is associated with some actual or potential environmental impact, which manifests itself in the form of physical, chemical, and biological disturbances in the water column, on the seabed, and in the atmosphere. Pollution occurs at all stages of offshore [exploration and production] … and involves more than 800 substances, dominated by oil and oil products.[32]

Historically, the greatest challenge faced by the *Arabian Gulf/Gulf of Oman Convention* was the impact of the 1991 Gulf War waged by Iraq upon Kuwait's petroleum-producing facilities – and its aftermath – which tested the provisions of this treaty to breaking point.[33] Since the Gulf

---

[30] Sergei Vinogradov, 'The Impact of the Deepwater Horizon: The Evolving International Legal Regime for Offshore Accidental Pollution Prevention, Preparedness, and Response' (2013) 44(4) *Ocean Development and International Law* 335–362.

[31] Simon Marsden, 'Regulatory Reform of Australia's Offshore Oil and Gas Sector after the Montara Commission of Inquiry: What about Transboundary Environmental Impact Assessment?' (2013) 15(1) *Flinders Law Journal* 41–53.

[32] Above n. 30, 337, footnotes deleted.

[33] Sheppard et al., above n. 18, 31; Katherine M Kelly, 'Declaring War on the Environment: The Failure of International Environmental Treaties During the Persian Gulf War' (1992) 7(4) *American University International Law Review* 921–950; ARG Price, 'Impact of the 1991 Gulf War on the Coastal Environment and Ecosystems: Current Status and Future Prospects' (1998) 24 (1–2) *Environment International* 91–96; Mahmoud Y Abdulraheem, 'War-related Damage to the Marine Environment in the ROPME Sea Area', in Jay E Austin and Carl E Bruch (eds), *The Environmental Consequences of War: Legal, Economic, and Scientific Perspectives* (Cambridge University Press, Cambridge, 2007) 338–352; and Al-Awadi Abdulrahman, *The Environmental Catastrophe in Kuwait* (ROPME, undated).

War, a considerable literature has developed in connection with international law, disasters and the environment.[34] It appears that, increasingly, the relationship between what is natural and what is made-made in this connection is no longer distinct. Examples abound, for example climate change,[35] and, as will be seen in Chapter 5, earthquakes.[36] In relation to oil pollution, this may occur in connection with conflict, but, as seen, may also be the result of offshore spills from installations or vessels. In connection with the latter, following on from the comments of Nadim et al.,[37] Stephens reports on the reason why accidental vessel source pollution incidents have been a major general concern in the following terms:

> Although the volume of pollutants entering the marine environment from casualties and collisions is smaller than that from the normal use of vessels, they have attracted more concern because of the severe and often irreparable impacts they have upon the local environments in which they occur. The magnitude of the risk posed by accidents at sea is mainly a function of the kind of material being shipped. Oil tankers and ships carrying noxious chemicals in bulk are appropriately the targets of the most attention.[38]

---

[34] See for example: Susan Breau, 'Protection of the Environment during Armed Conflict', in S Alam, MJH Bhuiyan, T Chowdhury and E Techera (eds) *Routledge Handbook of International Environmental Law* (Routledge, London, 2013) 617–632; Tim Stephens, 'Disasters, International Environmental Law and the Anthropocene', in Susan Breau and Katja LH Samuel (eds), *Research Handbook on Disasters and International Law* (Cheltenham, Edward Elgar, 2016) in press; Mahmoud Y Abdulraheem, 'War-related Damage to the Marine Environment in the ROPME Sea Area', in Jay E Austin and Carl E Bruch (eds), *The Environmental Consequences of War: Legal, Economic, and Scientific Perspectives* (Cambridge University Press, Cambridge, 2007) 338–352; and Samira AS Omar, Ernest Briskey, Raafat Misak and Adel ASO Asem, 'The Gulf War Impact on the Terrestrial Environment of Kuwait: An Overview', in Jay E Austin and Carl E Bruch (eds), *The Environmental Consequences of War: Legal, Economic, and Scientific Perspectives* (Cambridge University Press, Cambridge, 2007) 316–337.

[35] Z Lu, DG Streets, Q Zhang and S Wang, 'A Novel Back-Trajectory Analysis of the Origin of Black Carbon Transported to the Himalayans and Tibetan Plateau during 1996–2010' (2012) 39 *Geophysical Research Letters* LO1809.

[36] See for example: Malcolm Moore, 'Chinese Earthquake May Have Been Man-Made, Say Scientists', *The Telegraph*, 2 February 2009; and Richard A Lovett, 'Man-Made Earthquakes Cause Seismic Rumblings', *Cosmos Magazine*, 24 March 2014.

[37] Above n. 26.

[38] Above n. 34.

This particular issue has been largely addressed through implementation of Law of the Sea Convention (LOSC) obligations originally adopted in 1982.[39] Yet other concerns remain, however, including decommissioning of offshore installations,[40] and removal of sunken ships.[41] It was in recognition of the harm potential from oil spills generally in the Arabian Gulf/Gulf of Oman that in April 1978 the eight states of the subregion adopted the *Arabian Gulf/Gulf of Oman Convention* and associated *Action Plan*,[42] making the Gulfs one of the first UNEP Regional Seas.[43]

Khan and Al-Ajmi comment on the needs of the Gulfs following the conflict.[44] After highlighting two decades ago that the Gulf ecosystem is both naturally (due to being a shallow, semi-enclosed sea and hence a 'pollutant trap') and anthropogenically stressed (due to development), they emphasize a number of factors. These are: an integrated regional management strategy, national sustainable development policies, legalization of EIA processes and retrospective application, linking science with

---

[39] United Nations Convention on the Law of the Sea (Montego Bay, 10 December 1982), 1833 UNTS 3, in force 16 November 1994.

[40] R Desrina, Chairil Anwar and Tri Muji Susantoro, 'Environmental Impacts of the Oil and Gas Platform Decommissioning' (2013) 36 *Scientific Contributions Oil and Gas* 98; Youna Lyons, 'The New Offshore Oil and Gas Installation Abandonment Wave and the International Rules on Removal and Dumping' (2014) 29 *International Journal of Marine and Coastal Law* 498. See also 1989 Guidelines and Standards for the Removal of Offshore Installations and Structures on the Continental Shelf and in the Exclusive Economic Zone, IMO Resolution A. 672 (16) (19 October 1989).

[41] Nadim et al., above n. 26, 560–561.

[42] Action Plan for the Protection of the Marine Environment and the Coastal Areas of Bahrain, Iran, Iraq, Kuwait, Oman, Qatar, Saudi Arabia and the United Arab Emirates (Kuwait, 24 April 1978), unreported, in force 1 July 1979; <http://www.unep.org/regionalseas/programmes/nonunep/ropme/instruments/r_profile_ropme.pdf> accessed 7/3/16, 19–20.

[43] See Elizabeth Maruma Mrema, 'Regional Seas Programme: The Role Played by UNEP in its Development and Governance', in David Attard, Malgosia Fitzmaurice, Norman Martinez and Riyaz Hamza (eds), *The IMLI Manual on International Maritime Law. Vol III, Marine Environmental Law and International Maritime Security Law* (Oxford, Oxford University Press, 2016); and Nilufer Oral, 'Forty Years of the UNEP Regional Seas Programme: From Past to Future,' in Rosemary Rayfuse (ed), *Research Handbook on International Marine Environmental Law* (Edward Elgar, Cheltenham, 2015).

[44] Nuzrat Yar Khan and Dhari Al-Ajmi, 'Post War Imperatives for the Sustainable Management of the Gulf Eco-System' (1998) 24 (1–2) *Environment International* 239–248.

policy, and commitment to scientific research.[45] The Gulf (Kuwait) *Action Plan* (KAP) was intended to address many of these matters:

> The Kuwait Action Plan (KAP), to which all nations of the Gulf[s] are signatories, provides the basis for recognizing the Gulf[s] as an integrated and dynamic system and to develop a cooperative strategy for its sustainable management. The KAP, as part of UNEP's Regional Seas Program, attempts to provide a regional perspective to the protection and management of the Gulf ecosystem, and the Regional Organization for the Protection of the Marine Environment (ROPME), established under its mandate, provides the necessary focal point for regional cooperation required to carry out its objectives.[46]

The *Action Plan* mainly covers programme activities relating to oil pollution, industrial wastes, sewage and marine resources. It consists of four main components; environmental assessment; environmental management; a legal component; and institutional and financial arrangements. Protection of the marine environment is, however, the first priority, with a view to the promotion of human health and well-being as the ultimate goal. Khan and Al-Ajmi believe that 'the KAP provides an ideal set-up for regional cooperation in the sustainable management of the Gulf ecosystem',[47] although they emphasize the need for other regional organizations and agreements to be linked with ROPME's mandate under the KAP for greater effectiveness. The aims of the *Action Plan* are to achieve:

- Assessment of the state of the environment including socio-economic development activities related to environmental quality and of the needs of the Region in order to assist Governments to cope properly with environmental problems, particularly those concerning the marine environment;
- Development of guidelines for the management of those activities which have an impact on environmental quality or on the protection and use of renewable marine resources on a sustainable basis;
- Development of legal instruments providing the legal basis for cooperative efforts to protect and develop the Region on a sustainable basis; and

---

45 Ibid, 243–247.
46 Ibid, 243.
47 Ibid, 244.

- Supporting measures including national and regional institutional mechanisms and structure needed for the successful implementation of the Action Plan.[48]

Projects range over coastal area management, fisheries, public health, land-based activities, sea-based pollution, biodiversity, oceanography, marine emergencies, Geographic Information Systems (GIS) and remote sensing, environmental awareness and capacity building. Milestones included the creation in 1979 of the Regional Organization for the Protection of the Marine Environment (ROPME), the establishment in 1982 of the Marine Emergency Mutual Aid Centre (MEMAC), and the adoption of four protocols, the first of which established MEMAC.[49] These are the *Protocol Concerning Regional Cooperation in Combating Pollution by Oil and Other Harmful Substances in Cases of Emergency* (*'Regional Cooperation Protocol'*),[50] the *Protocol Concerning Marine Pollution Resulting from Exploration and Exploitation of the Continental Shelf* (*'Continental Shelf Protocol'*),[51] the *Protocol for the Protection of the Marine Environment against Pollution from Land Based Sources* (*'Land-Based Sources Protocol'*),[52] and the *Protocol on the Control of Marine Transboundary Movements and Disposal of Hazardous Wastes and other Wastes* (*'Hazardous Wastes Protocol'*).[53] There is also an unadopted fifth *Protocol Concerning the Conservation of Biological Diversity and the Establishment of Protected Area* (*'Biodiversity Protocol'*), although at the time of writing (2016) a draft of this is not publicly available.[54]

---

48 Above n. 42, 20.

49 MEMAC is also established under the Red Sea/Gulf of Aden Convention and various protocols, as indicated in the text below.

50 ROPME, Protocol Concerning Regional Cooperation in Combating Pollution by Oil and Other Harmful Substances in Cases of Emergency (Kuwait, 24 April 1978), unreported, in force 20 August 1985.

51 ROPME, Protocol Concerning Marine Pollution Resulting from Exploration and Exploitation of the Continental Shelf (Kuwait, 29 March 1989), unreported, in force 17 February 1990.

52 ROPME, Protocol for the Protection of the Marine Environment against Pollution from Land-Based Sources (Kuwait, 21 February 1990).

53 ROPME, Protocol on the Control of Marine Transboundary Movements and Disposal of Hazardous Wastes and other Wastes (Tehran, 17 March 1998), Ecolex, B7 p. 978:31/D, in force 21 August 2005.

54 This has more recent resonance given the adoption of the Biodiversity Convention since the adoption of the Arabian Gulf/Gulf of Oman, and the Red

ROPME is established under Article XVI of the Convention, with its headquarters in Kuwait. Nadim et al. comment that 'ROPME is now the most suitable basis for future cooperation between the Gulf littoral states regarding various aspects of this semi-enclosed marine environment'.[55] It consists of a decision-making Council constituted by the member states, an administering Secretariat, and a Judicial Commission for the Settlement of Disputes. In connection with the latter in particular, Mahmoudi comments:

> The Kuwait Convention is ... different from most of the other UNEP conventions with respect to its institution and its dispute settlement procedure. Instead of following the traditional pattern of conducting business through the meetings of the Contracting Parties, the Convention, in Article XVI, provides for the establishment of a Regional Organization for the Protection of the Marine Environment (ROPME) ... The UNEP regional seas conventions usually provide for arbitration as the method of dispute settlement if the parties cannot solve their problems through negotiations. The dispute settlement procedure in the Kuwait Convention is different in that a specific Judicial Commission for that purpose is established. The composition, terms of reference and rules of procedure of the Commission are established by the Council. In this way, a predetermined legal forum for dispute settlement is created which, unlike arbitration with its *ad hoc* nature, can give a higher sense of legal security.[56]

The provisions relating to the Council are outlined in Article XVII, detailing the requirements for meetings, chairmanship and voting procedure in Article XVII (a)–(c). The numerous Council functions are described in Article XVII(d)–(x), and include: implementation review of the Convention and Action Plan; review of the state of marine pollution and its effects in the Sea Area; adoption, review and amendment of annexes to the agreements; receiving and considering reports by the member states and the Secretariat; making recommendations concerning additional protocols or amendments;[57] establishing subsidiary bodies, and other matters. Article XVIII concerns the functions of the Secretariat,

---

Sea/Gulf of Aden Conventions. See: Convention on Biological Diversity (CBD) (Rio de Janiero, 22 May 1992) 31 ILM 822, in force 29 December 1993.

[55] Above n. 26, 559.

[56] Said Mahmoudi, 'Legal Protection of the Persian Gulf's Marine Environment' (1997) 21(1) *Marine Policy* 53, 55–56.

[57] Provided for in accordance with Articles XIX, XX and XXI, Arabian Gulf/Gulf of Oman Convention, above n 1.

which include considering enquiries, the preparation of reports, maintaining a database of national laws, and providing technical assistance and training programmes.

The strengths, weaknesses and potential of ROPME is analysed by Nadim et al. in the following terms.[58] In relation to strengths, they point to the 'vital common goals' found within the regime which have provided 'the ground for close cooperation'. Furthermore, and relating to one of the factors emphasized by Khan and Al-Ajmi above, 'Links between scientists and policy makers within the framework of ROPME have created a unique collaborative atmosphere among the littoral States leading them to sound environmental decisions'.[59] Meetings between the state Parties, and collaboration between ROPME and other regional organizations (again emphasized by Khan and Al-Ajmi above), have also been advantageous.

The weaknesses found relate to inadequate implementation and compliance, in particular: lack of prioritization of critical coastal issues, uncoordinated coastal zone management and among government institutions, inadequate funding and information exchange, inadequate and insufficient operational and management planning and human resourcing, the potential to withdraw from the regime, and suspension of ROPME programmes due to regional conflicts in the past three decades.[60] The potential of ROPME is nonetheless emphasized in a number of areas, including establishing a trust fund and data bank, creating universal discharge standards and a pollution credit scheme.[61]

2.2 Objectives, Principles and Procedures, including Implementation and Compliance

2.2.1 Convention

There is no specifically stated objective or objectives of the *Arabian Gulf/Gulf of Oman Convention*. However, the title and recitals leave the reader in no doubt as to its purpose. The title makes clear that the focus is squarely upon the marine environment and the prevention of pollution based on cooperation between the member states, with the recitals emphasizing that 'the States sharing the Region have a special responsibility to protect its marine environment'.[62] 'Oil and other harmful or

[58] Nadim et al., above n. 26, 562–563.
[59] Nadim et al., above n. 26, 562.
[60] Nadim et al., above n. 26, 562.
[61] Nadim et al., above n. 26, 563.
[62] Arabian Gulf/Gulf of Oman Convention, above n. 1, recital 7.

noxious materials arising from human activities on land or at sea'[63] are the overriding concern, which are derived from 'urban and rural development',[64] and 'processes of industrial development'.[65] Regulating discharges is particularly important in these contexts, with ROPME's task emphasized:

> With the health of the population living along the shores of the Gulf and utilizing its desalinated waters for drinking purposes at stake, it becomes a vital task for ROPME to take the leadership and oversee the treatment and enforcement of discharge regulations of domestic and industrial wastewater discharges into the Persian Gulf. Prime attention should be given to discharges originating from refineries and petrochemical companies that contain large quantities of toxic wastes.[66]

An integrated management approach is required under the Convention, to be supported by scientific research and a monitoring and assessment programme.[67] ROPME itself sets out the objectives of the Convention as follows:

> The main objective of ROPME is to coordinate efforts of the eight Member States towards protection [of] the marine and coastal environment and ecosystems in the ROPME Sea Area against marine pollution and stressors that might be induced from developmental activities or/and other drivers of change.[68]

Article 1 contains the definition section, describing 'marine pollution', 'national authority', 'organization', 'secretariat' and 'Action Plan'; Article 2 outlines the geographical coverage of the treaty, which is referred to as the 'Sea Area'. This includes both the Arabian Gulf and Gulf of Oman, although the former is frequently used as a term to include both. Article 3 outlines the general obligations. Article 3(a) therefore requires the Contracting States to 'prevent, abate and combat pollution'; Article 3(b) obliges them to 'co-operate in the formulation and adoption of other protocols prescribing agreed measures, procedures and standards', which indicates the framework nature of the *Arabian Gulf/Gulf of Oman Convention*; Article 3(c) emphasizes that these standards – as well as laws and regulations – are, while national, to be harmonized; Article 3(d)

---

63 Ibid, recital 1.
64 Ibid, recital 3.
65 Ibid, recital 4.
66 Nadim et al., above n. 26, 559.
67 Arabian Gulf/Gulf of Oman Convention, above n. 1, recital 6.
68 <http://ropme.org/1_WhoWeAre_EN.clx#> accessed 14/11/16.

furthermore requires cooperation with 'the competent international, regional and sub-regional organizations' in relation to regional standards – which suggest UNEP in particular and the Regional Seas Programme – and finally, Article 3(e) emphasizes the need to avoid 'transformation of one type of pollution to another which could be more detrimental to the environment'.

Articles IV–VIII concern different types of pollution, or rather those with different origins. Article IV is hence concerned with pollution from ships, whether 'intentional or accidental', and requires conformity between the *Arabian Gulf/Gulf of Oman Convention* and 'the applicable rules of international law'. These types of incident were examined above; clearly regulation must be coordinated, with both treaty and general international law applicable, in particular the LOSC.[69] Article V addresses dumping of waste from ships and aircraft, again with reference to 'relevant international conventions', which has specific reference to the IMO *Convention on the Prevention of Marine Pollution by Dumping of Wastes and Other Matter*.[70]

Article VI concerns pollution from land-based sources, 'whether water-borne, air-borne, or directly from the coast including outfalls and pipelines', which, as indicated in the recitals discussed above, may come from urban, rural or industrial development. Article VII is focused on pollution from exploration and exploitation of the seabed and continental shelf, with requirements to prevent, abate and combat pollution extending to accidents and emergencies. Finally, Article VIII, while misleadingly entitled 'pollution from human activities' – which suggests urban, rural and industrial development broadly – is in fact concerned specifically with land reclamation and related dredging.

The next two obligations in the treaty address cooperation between the member states: Article IX pollution emergencies, and Article X scientific and technological cooperation. In relation to the first, there are requirements to ensure adequate equipment and qualified personnel are available

---

[69] Above n. 39. See also: Robin Warner, 'Conserving Marine Biodiversity in the Global Marine Commons: Co-Evolution and Interaction with the Law of the Sea' (2014) 6(1) *Frontiers in Marine Science* 1–23.

[70] Convention on the Prevention of Marine Pollution by Dumping of Wastes and Other Matter (Washington DC, Moscow, London, Mexico City, 29 December 1972) 1046 UNTS 120, in force 30 August 1975. For discussion, see Hossein Esmaeili and Brendan Grigg, 'Pollution from Dumping', in David Attard, Malgosia Fitzmaurice, Norman Martinez and Riyaz Hamza (eds), *The IMLI Manual on International Maritime Law. Vol III, Marine Environmental Law and International Maritime Security Law* (Oxford University Press, Oxford, 2016).

to deal with emergencies and to reduce or eliminate damage from them; any member state becoming aware of such emergency is obliged to notify ROPME, and via the Secretariat, any Party likely to be affected. In regard to the second, scientific and technological cooperation is required. This is needed directly between the member states, or indirectly via 'competent international and regional organizations' – which may include UNEP and the IMO – in respect of research, monitoring and assessment. While this includes data exchange generally in Article X(a), more specifically the obligations in Article X(b) extend to the development and coordination of national research and monitoring programmes, and to the establishment of a regional network in cooperation with the likely competent regional or international organizations indicated.

Provisions for environmental impact assessment (EIA) are contained in Article XI. Article XI(a) requires Parties to 'include an assessment of the potential environmental effects in any planning activity entailing projects within its territory … which may cause significant risks of pollution'. Despite the clumsy language, the date of adoption of the Convention indicates that there is no intention for an EIA of plans to take place, but rather specific projects. Article XI(b) allows for procedures to be elaborated in relation to information to be released in relation to the assessment; and in accordance with Article XI(c), Parties undertake to develop guidelines 'to assist the planning of their development projects … to minimize their harmful impact on the marine environment'. Examples from Bahrain can be found in the literature, where domestic EIA law has been criticized. Naser et al. comment for example: 'Since the legal requirements for EIA may not guarantee that biodiversity and ecological issues are taken into account, consideration should be given to integrating such issues into existing environmental law in Bahrain'.[71] Note also that another MEA, the *World Heritage Convention*,[72] has a Marine Programme which has produced a 'Bahrain' Action Plan for general application; EIA is commonly used in relation to proposals that may impact World Heritage.[73]

---

[71] Humood Naser, John Bythell and Jeremy Thomason, 'Ecological Assessment: An Initial Evaluation of the Ecological Input in Environmental Assessment Reports in Bahrain' (2008) 26(3) *Impact Assessment and Project Appraisal* 201, 206.

[72] Convention Concerning the Protection of the World Cultural and Natural Heritage (Paris, 16 November 1972) 11 ILM 1358, in force 17 December 1975.

[73] D Laffoley and J Langley (eds), *The Bahrain Action Plan for Marine World Heritage: Identifying Priorities for Marine World Heritage and Enhancing the Role of the World Heritage Convention in the IUCN WCPA Marine Global*

Khan and Al-Ajmi also emphasize not only the role of EIA, but also strategic environmental assessment (SEA). This process applies EIA to policies, plans and programmes.[74] They believe this 'can play a significant role in the sustainable management of the Arabian Gulf', and

> can be an impetus for the administrative change needed … [which] would be consistent with the objectives of the KAP … One of its most significant contributions would be the enhancement of interagency coordination and dissemination of information about the proposed actions (policies, programs, and projects) and their biophysical and socioeconomic impacts which is prominently lacking in the Arabian Gulf region.[75]

As indicated above, the Gulf War of 1991 put the Arabian Gulf/Gulf of Oman regime under considerable pressure, and the environmental catastrophe in Kuwait in particular was reported on at length.[76] Yet despite the serious consequences of the Gulf War and other recent conflicts, such as in Yemen,[77] in more recent times it is the pace and scale of development that is the more practical concern in many states in the subregion; in particular, the cumulative significant negative effects upon the environment which have been and are increasingly beyond the carrying capacity of the Gulf states:

> Most scientists working in the Gulf agree that the most significant and important threats to the sustainability of the Gulf ecosystems come from the massive extent of coastal habitat modification by dredging and converting shallow, productive marine areas into land for homes, recreation and industrial facilities … Many projects involve massive deposition of material into shallow waters, resulting in replacement of several square km of productive tidal flat by inert fill material. Creation of offshore islands and structures has involved destructive dredging of seagrass and algal beds whose 'reclaimed'

Plan of Action for MPAs in Our Oceans and Seas (IUCN, Gland, 2010). In relation to SEA for example, see Simon Marsden: 'Protecting Heritage on Australia's Coasts: A Role for Strategic Environmental Assessment?' (2013) 15(3) Journal *of Environmental Assessment Policy and Management* 1350014-1–1350014-21.

[74] See Simon Marsden, *Strategic Environmental Assessment in International and European Law* (Earthscan, Abingdon and New York, 2008).

[75] Khan and Al-Ajmi above n. 44, 246.

[76] See references above n. 33; also Muhammad Sadiq and John Charles McCain, *The Gulf War Aftermath: An Environmental Tragedy* (Springer, 1993).

[77] See for example: AFP, 'US Military Admits Troops Are Operating Inside Yemen to Combat Al-Qaida', *The Guardian*, 6 May 2016.

material, taken from 'borrow pits' then further destroys other areas onto which it is dumped.[78]

In relation to project level EIA, although this is an obligation of the Convention, it has been found wanting in practice, suggesting there is need for other approaches such as an SEA to be utilized in advance. Sheppard et al. comment:

> Most projects include 'baseline' surveys using a range of tools such as environmental impact assessments (EIA), in which, typically, the existing habitat is examined and a judgement made on the extent of potential impacts, and proposals on how to reduce these and monitor them. Sometimes expensive and lengthy modelling is undertaken as part of engineering studies. But commonly, the science has the ability to advise only slight changes to a project, and what is often missed also is that the 'baseline' chosen is little more than an already severely degraded marine ecosystem. Today it is difficult to find any meaningful baselines, not only because of ongoing, intensive constructions, but also because of several recent episodes of marine mortality from seawater warming.[79]

In relation to liability and compensation for damage resulting from marine pollution, Article XIII requires Parties to 'co-operate in the formulation and adoption of appropriate rules and procedures' which are focused on civil claims, and must also have regard to 'international rules and procedures'.[80] Exemptions are provided in Article XIV from the

---

[78] Sheppard et al., above n. 18, 25. Large-scale dredging and reclamation activities are particularly problematic; see in other parts of Asia, for example: Dae-In Lee, Ki-Hyuk Eoma, Gui-Young Kim and Gun-Wook Baeck, 'Scoping the Effective Marine Environmental Assessment of Dredging and Ocean Disposal of Coastal Sediments in Korea' (2010) 34 *Marine Policy* 1082–1092; and Yin-Ki Tam, I-Hsun Ni, Cynthia Yau, Man-Yee Yan, Wai-Shan Chan, Sze-Man Chan and Hsueh-Jung Lu, 'Tracking the Changes of a Fish Community Following a Megascale Reclamation and Ensuing Mitigation Measures' (2013) 70(6) *ICES Journal of Marine Science* 1206–1219. Note also the recent findings of the PCA in relation to Chinese reclamation in the South China Sea, which is discussed in Chapter 5 in particular.

[79] Sheppard et al., above n. 18.

[80] See for example: Amy Aai Sheau Ye, 'Liability and Compensation Regime for Transboundary Oil Pollution Damage' (2014) 5(2) *Australian Journal of Maritime and Ocean Affairs* 59–64; Kissi Agyebeng, 'Disappearing Acts – Towards a Global Civil Liability Regime for Pollution Damage Resulting from Offshore Oil and Gas Exploration' (2006) *Cornell Law School Graduate Student Papers*, Paper 11; and Mans Jacobsson, 'Liability and Compensation for Ship Source Pollution', in David Attard, Malgosia Fitzmaurice, Norman Martinez and

obligations of the Convention and its protocols in connection with state vessels including warships, although in each case it is expected that the operation of such craft will comply 'as far as possible' with the requirements of the agreement.

Provisions for national implementation reporting (NIR) and compliance control are found in Articles XXIII, which concerns the NIR mechanism, and XXIV, which concerns compliance control. Article XXIII requires Contracting States to submit reports to the Secretariat outlining their implementation efforts in accordance with instructions from the Council. Article XXIV requires Contracting States to cooperate in developing compliance procedures for detection of violations and environmental monitoring. Dispute settlement provisions are found in Article XXV and oblige in the first instance peaceful resolution, failing which disputes are to be submitted to the Judicial Commission for the Settlement of Disputes. Provisions in the *Regional Cooperation Protocol*, *Continental Shelf Protocol*, *Land-Based Sources Protocol*, and *Hazardous Wastes Protocol* are analysed in the next section, together with the draft *Biological Diversity Protocol.*

2.2.2 Protocols

The protocols to the *Arabian Gulf/Gulf of Oman Convention* contain more detailed obligations than the Convention itself, which also apply to different territory. Mahmoudi explains and clarifies this as follows:

> As is the case with most of the environmental conventions, the detailed substantive rules should be sought in the protocols to the Kuwait Convention. The areas of application of these protocols differ to some extent. The 1989 [Continental Shelf] Protocol applies to the Sea Area as defined in the Convention, i.e., the territorial seas and the marine areas beyond them. The 1978 [Regional Cooperation] Protocol applies in the Convention area as is defined in Article II of the Convention. It can be applied even in internal waters if the concerned State Party so permits in the event of an emergency. The 1990 Protocol on land-based pollution applies, in addition to the Convention area, to the waters on the landward side of the baselines, and in the case of watercourses, up to the freshwater limit.[81]

In relation to the earliest of these agreements – and first outlined here – the *Regional Cooperation Protocol* addresses the urgency of responding

---

Riyaz Hamza (eds), *The IMLI Manual on International Maritime Law. Vol III, Marine Environmental Law and International Maritime Security Law* (Oxford University Press, Oxford, 2016).

[81] Above n. 56, 56, footnotes deleted.

to pollution emergencies by establishing cooperative, comprehensive and effective measures. As indicated earlier in this chapter, and as discussed in the practice examples below, this was an essential part of the response to the damage caused by the Gulf War. Article II of this Protocol outlines the objectives to: 'co-operate in taking the necessary and effective measures to protect the coastline and related interests' and 'to maintain and promote … their contingency plans and means for combating pollution'. Article I contains key definitions of 'Appropriate Authority', 'Marine Emergency', 'Marine Emergency Contingency Plan', 'Marine Emergency Response' and other matters. Marine Emergency for example is defined in Article I(2) as meaning:

> … any casualty, incident, occurrence or situation, however caused, resulting in substantial pollution or imminent threat of substantial pollution to the marine environment by oil or other harmful substances and includes, *inter alia,* collisions, strandings and other incident involving ships, including tankers, blow-outs arising from petroleum drilling and production activities, and the presence of oil or other harmful substances arising from the failure of industrial installations.

The purpose of the Contingency Plan is to 'co-ordinate the deployment, allocation and use of personnel, material and equipment' in such situations (Article I(3)); with the 'Response' being the activities 'to prevent, mitigate or eliminate' the pollution or threat thereof (Article I(4)). 'Related Interests' of the affected state are furthermore defined in Article I(5) as including directly or indirectly affected or threatened activities, attractions, or public health. Of the operational provisions, Article III establishes MEMAC with the objectives of coordinating efforts, developing national capacity and initiating operations to combat pollution (Article III(2)). Mahmoudi describes MEMAC and explains its role as: 'the operational unit of ROPME [which] acts according to the guidelines decided by the Council of ROPME. However, in practice it functions more or less independently from ROPME'.[82] Specific functions include collating and circulating information; assisting the Contracting States in relation to the preparation of laws, plans and procedures; coordinating training programmes; developing and maintaining an emergency information system; preparing inventories of available personnel, material vessels, etc.; establishing and maintaining liaison with regional and international organizations; and preparing periodic reports on emergencies for the Council (Article III(3)).

---

[82] Above n. 56, 56.

Other provisions in the *Regional Cooperation Protocol* include the following: provision of information on appropriate authorities, applicable laws and contingency plans from Contracting States to the Centre and other Contracting States (Article V); provision of information on marine emergency responses (Article VI); obligations to require operators to report marine emergencies (Article VII);[83] requirements to the Centre to transmit information received from Contracting States to all other Contracting States (Article VIII); obligations on Contracting States to take appropriate measures to combat pollution, inform others and assess the nature and extent of the emergency and action required (Article X); provisions relating to requests for assistance (Article XI); and requirements on appropriate authorities of states to fully carry out their obligations (Article XII).

Regarding the second of these agreements, the *Continental Shelf Protocol*, while not stated explicitly, the objective is 'to prevent and control marine pollution from exploration and exploitation of the sea-bed and its subsoil' (Recital 4). It has been described as 'the first comprehensive international agreement regulating in detail pollution caused by exploitation of the sea-bed'.[84] After the definition section in Article I, Article II of the Protocol therefore obliges Contracting States to take 'all appropriate measures' to 'prevent, abate and control' marine pollution from offshore operations; Article III requires such operations to be licensed subject to conditions protecting 'the marine environment and coastal areas'; and before licensing any such operation the competent authority must call for 'submission of an assessment of the potential environmental effects' (Article IV(1a)). This EIA provision relates to Article XI of the Convention, and applies here specifically to offshore operations. The assessment is to be conducted in accordance with the ROPME Guidelines.[85] These contain provisions on the functions of the authorities, procedures, criteria in parts A–C and two annexes containing a matrix and detailed criteria for assessing the effects on flora and fauna. If an assessment is not called for, a survey of the marine environment is suggested as an alternative.

---

[83] These reports are to conform with the guidelines in Appendix A to the Protocol.

[84] Mahmoudi, above n. 56, 56.

[85] ROPME, *Guidelines on Requirements for the Conduct of Environmental Impact Surveys and the Production of Environmental Impact Statements*, MEMAC undated, <http://www.memac-rsa.org/forms_and_downloads> accessed 12/5/15.

Other provisions in the *Continental Shelf Protocol* relate to the avoidance of interference with lawful navigation, fishing or other authorized activities (Article V); ensuring the safety and fitness for purpose of offshore installations (Articles VI and VII); the presence and approval of contingency plans by the competent authority for offshore operations (Article VIII); the need for more stringent controls in relation to any 'Special Area' (Article IX),[86] (which include the need for treatment and appropriate disposal of drill cuttings where oil-based fluids are used in these areas);[87] the establishment of practicable measures to control disposal at sea (Article X); appropriate measures for a Chemical Use Plan, regulation of chemicals,[88] and seismic operations in accordance with ROPME Guidelines (Article XI);[89] and effective collection, disposal and use of unwanted substances or articles in offshore operations (Article XII). Additional regulatory measures concern the need for Competent State Authorities to have power over pipelines and platforms (Article XIII(1)), stipulating that states adopt a common policy on decommissioning installations where fishing grounds are concerned (Article XIII(2)) and avoid depositing unwanted installations on the sea-bed (Article XIII(3)). Decommissioning of disused offshore installations has been an issue for some time, as indicated early in this chapter. Mahmoudi comments on this provision:

---

[86] The ROPME Sea Area which encompasses both the Arabian Gulf and Gulf of Oman became a Special Area in accordance with the IMO Resolution as from 1 August 2008, and is therefore subject to the discharge requirements of Regulation 1.11.5 of Annex I and Regulation 5(1)(e) of Annex V of the MARPOL 73/78 Convention. See International Convention for the Prevention of Pollution from Ships (London, 11 February 1973) 12 ILM 1319 (1973), in force 19 February 1983. Resolution MEPC, 168(56) adopted on 13 July 2007, Establishment of the date on which the amendments to Regulation 1.11.5 of MARPOL Annex I and amendments to Regulation 5(1)(e) of MARPOL Annex V in respect of the Gulf's Special Area shall take effect.

[87] ROPME, *Guidelines on the Voluntary Regulation of the Disposal of Drill Cuttings on the Sea Bed*, MEMAC, undated, <http://www.memac-rsa.org/sites/default/files/formsAndDownloads/4%20Guidelines%20on%20the%20voluntary%20regulation%20of%20the%20disposal%20of%20drill%20cuttings%20on%20the%20sea%20bed.pdf> accessed 7/3/16.

[88] ROPME, *Guidelines on the Use and Storage of Chemicals in Offshore Operations*, MEMAC, undated. <http://www.memac-rsa.org/forms_and_downloads> accessed 12/5/15.

[89] ROPME, *Guidelines on the Conduct of Seismic Operations*, MEMAC, undated, <http://www.memac-rsa.org/forms_and_downloads> accessed 12/5/15.

> Article XIII of the Protocol addresses the issue of the removal of abandoned and disused offshore installations. Due to the aging of many offshore platforms in the middle of the 1980s, this was a topical environmental issue. The adoption of the 1989 Protocol was preceded by the debates in IMO meetings concerning decommissioned offshore installations.[90]

Comparisons between the UNEP Regional Seas arrangements governed by international agreements are a helpful means of highlighting both differences and regulatory failure. Vinogradov for example compares the *Continental Shelf Protocol* with more recent arrangements elsewhere, in particular under *Mediterranean (Madrid) Protocol*[91] as follows:

> Both instruments cover the entire range of the environmental aspects of offshore E&P, from construction authorization and prevention of operational and accidental pollution to decommissioning. They include regulations specifically tailored to address pollution emergency response. Both protocols have provisions on safety, contingency planning, notification, and assistance.[92]

However there is a clear difference between the measures in the two protocols identified, which leaves the arrangements for the Gulfs found wanting:

> Compared to the earlier Kuwait Protocol, the Madrid Protocol's two annexes (Annex VI 'Safety Measures' and Annex VII 'Contingency Plan') are more detailed such that the regime for the accidental pollution prevention, preparedness, and response in the Mediterranean Sea is the most developed among similar arrangements. It is based on a broad and comprehensive normative framework prescribing clear instructions to both government authorities and the oil industry and is supported by an appropriate institutional mechanism.[93]

---

90 Above n. 56, 57.

91 Protocol for the Protection of the Mediterranean Sea Against Pollution Resulting from Exploration and Exploitation of the Continental Shelf and the Seabed and its Subsoil (Madrid Offshore Protocol), 14 October 1994, available at <faolex.fao.org/docs/pdf/mul38165.pdf> accessed 18/05/16. The protocol entered into force on 24 March 2011. Note that this is a protocol under the Convention for the Protection of the Marine Environment and the Coastal Region of the Mediterranean (Barcelona, 10 June 1995) 1102 UNTS 27, in force 9 July 2004; significantly this 'Barcelona Convention' shares the same feature of the Red Sea/Gulf of Aden Convention in also being a transcontinental regime, in the former case however combining state Parties from the three continents of Europe, Africa and Asia – the latter states are Israel, Syria and Turkey.

92 Above n. 30, 346.

93 Above n. 30, 346.

Comparisons between the UNEP Regional Seas arrangements may in the future be made easier also in connection with evaluating effectiveness. Based on the ecosystem approach to environmental management – where the integrity of the ecosystem is prioritized over individual species or site protections – comprehensive indicators may enable comparisons to be made between the governance measures more effectively. A draft set of coordinated indicators has recently been produced by UNEP,[94] which it is hoped will link with the UNDP Sustainable Development Goals (SDGs).[95] The SDGs build on the previous Millennium Development Goals (MDGs), broadening them within the sustainability agenda, and in so doing, 'addressing the root causes of poverty and the universal need for development that works for all people'.[96]

The third of the related agreements is the *Land-Based Sources Protocol.* The primary objective of this is implicit from Recital 2, which notes the 'danger posed to the marine environment' from 'the release of untreated, insufficiently treated and/or inadequately disposed of domestic or industrial discharges'. The sources of pollution are indicated in more detail in Article III, with the focus of efforts including pollution from 'outfalls and pipelines discharging into the sea'; and 'through rivers, canals and other watercourses'. Appendix I to the Protocol explains the approach of pollution abatement through source control which underlines the requirements of Article IV; this requires implementing action programmes, associated measures and timetables for implementation. The Protocol emphasizes the collective location of industrial facilities to enable joint, and/or combined effluent treatment, which is the focus of Article V and Annex II.

In accordance with Article VI, guidelines, regulations and permits for the release of wastes are to be based on Annex III of the Protocol. Regional guidelines are to be used to determine the quality of sea water, and regional regulations for waste discharge and/or treatment. The weakness of this protocol is essentially that 'it is evident from the relevant articles that a certain amount of land-based pollution for the sake of industrialization should be tolerated'.[97] It is however acknowledged

---

94 David Johnson, Angela Benn and Maria Adelaide Ferreira, *Measuring Success: Indicators for the Regional Seas Conventions and Action Plans* (United Nations Environment Programme, 2014). See pages 22–24 for details about each of the Regional Seas Programmes.

95 <http://www.undp.org/content/undp/en/home/sdgoverview/post-2015-development-agenda.html> accessed 22/6/16.

96 Ibid.

97 Mahmoudi, above n. 56, 57.

that stricter regulations may be needed in specific instances in accordance with the 'local ecological, geographical and physical characteristic' and the 'level of existing pollution' (Article VI(1)). Cost implications as part of a sustainable development approach are expected to inform such programmes (Article VI(2)). Guidelines, standards and criteria must be updated regularly (Article VI(4)), and fed into the review and modification of permit conditions of polluters (Article VI(5)). Provisions for monitoring and data management are outlined in Article VII and EIA in Article VIII. Of the latter, which again relates to Article XI of the Convention, this is expected 'on a priority basis' in coastal areas where there are significant risks of pollution from land-based activities (Article VIII(1)). Technical and other guidelines are to be developed for assessment, including of transboundary effects (Article VIII(2)). The implementation of any project following an EIA must take 'fully into account the findings' (Article VIII(3)).

Other provisions in the *Land-Based Sources Protocol* concern scientific and technological cooperation and assistance (Articles IX and X); watercourses shared by states (Article XI); exchange of information (Article XII); and responsibility and liability for damage (Article XIII). In relation to shared watercourses, where pollution originates in a non-Contracting State, the affected Contracting State is urged to cooperate with that state in applying the Protocol fully, although is not to be held responsible for the pollution (Article XI(2)). Where however damage is caused to a Contracting State by another Contracting State, 'prompt and adequate compensation or other relief' is required (Article XIII(1)), in accordance with procedures to be formulated and adopted under the Protocol (Article XIII(2)).

A significant general problem identified by Kirk in connection with land-based sources of pollution is the lack of precision concerning obligations, which makes evaluating compliance extremely difficult. She analyses another protocol under the UNEP Regional Seas Programme for the Baltic, known as the *Helsinki Convention*,[98] making the following observations of relevance to the Kuwait arrangements:

> … the Helsinki Convention requires parties 'to prevent and eliminate pollution of the Baltic Sea Area from land-based sources by using, inter alia, Best Environmental Practice for all sources and Best Available Technology for point sources.' This action-oriented provision is elaborated on in Annex III of

[98] Convention on the Protection of the Marine Environment of the Baltic Sea Area (Helsinki, 9 April 1992) 1507 UNTS 167, in force 17 January 2000. See <http://www.helcom.fi> accessed 12/5/16.

> the Convention; however, particularly where it relates to best environmental practices, Annex III remains imprecise, again providing an action-oriented approach.[99]

The fourth subsidiary agreement is the *Hazardous Wastes Protocol.* The objective of this is, broadly speaking, regional cooperation and coordination to control the movement of these wastes among Contracting States, and the imposition of restrictions on the importation of such wastes from non-Contracting States for disposal (Recital 5). The Protocol acknowledges the place of relevant MEAs in setting the context,[100] as well as the maritime agreements LOSC, MARPOL,[101] and the *London Dumping Convention.*[102] The scope of the Protocol in Article 1 applies to wastes appearing in its various Annexes; other wastes are excluded because they are radioactive or derive from offshore installations (which are regulated by other international agreements or the Kuwait *Continental Shelf Protocol* as appropriate).

General obligations are outlined in Article 4 after a definition section (Article 2) and area of application section (Article 3). These include that generation of hazardous wastes are to be kept to a minimum (Article 4(1)); that transportation and disposal are prohibited unless authorized (Article 4(2)); that appropriate measures are taken to avoid pollution or minimize its consequences (Article 4(3)); and that where hazardous wastes are moved across borders they be managed in accordance with international rules and standards (Article 4(4)). Specific provisions on importation and disposal (Articles 5 and 6); exportation to non-Contracting States (Article 7); movement between Contracting States (Article 8); third party use of port facilities (Article 9); illegal traffic (Article 10); and the duty to re-import (Article 11) follow.

---

[99] Above n. 20, 240.

[100] In particular, the Convention on the Control of Transboundary Movements of Hazardous Wastes and their Disposal (Basel, 22 March 1989) 1673 UNTS 126, in force 5 May 1992.

[101] International Convention for the Prevention of Pollution from Ships, 1973 (above n. 86), as modified by the Protocol of 1978 (London, 17 February 1978) 1340 UNTS 61, in force 2 October 1983.

[102] Above n. 70.

3. *RED SEA/GULF OF ADEN CONVENTION* AND ASSOCIATED PROTOCOLS

3.1 Geographical Context, Historical Background and Institutional Arrangements

The seven state Parties to the *Red Sea/Gulf of Aden Convention* are: Djibouti, Egypt, Jordan, Saudi Arabia, Somalia, Sudan and Yemen; only three of these are in southwest Asia, with the others in north Africa, making this a transcontinental subregional regime. The Red Sea and Gulf of Aden are recognized for their coastal and marine environments as well as related strategic, economic and social values. The increase in human population and rapid economic growth has put considerable pressure on the environment, with negative effects including marine pollution, depletion of marine resources and overfishing.[103] The fact that these are in many instances transboundary requires regional cooperation, which led to the initiation in 1974 of the Programme for the Environment of the Red Sea and Gulf of Aden (PERSGA).[104] This was in collaboration with the Arab League Educational, Cultural and Scientific Organization (ALECSO), and with the support of UNEP as one of its Regional Seas Programmes.[105] UNEP comments on the challenges facing the subregion as follows:

> Rapid coastal development associated with growing human usage of coastal/marine resources, and the ever rising maritime activities through the Red Sea and Gulf of Aden, as a main east-west maritime trading route, constitute the main challenges to protection and conservation of marine environment in the region.
>
> Major threats to the environment and coastal/marine resources include habitat destruction; non-sustainable use of living marine resources; navigation risks and risks from petroleum production and transport; urban and industrial hotspots; and rapid expansion of coastal tourism.[106]

---

[103] See Ruth Lapidoth-Eschelbacher, *The Red Sea and the Gulf of Aden* (Martinus Nijhoff, The Hague, 1982) and generally, Don Hinrichsen, above n. 23, Chapter 15, 'The Red Sea and the Gulf of Aden: Between Two Deserts'.

[104] PERSGA, Cairo Declaration Concerning the Formation of the Regional Organisation for the Conservation of the Environment of the Red Sea and Gulf of Aden, PERSGA, Council Decision No. (CD-1-GM1-26/9/1995).

[105] <http://www.unep.org/regionalseas/programmes/nonunep/redsea/> accessed 19/3/15.

[106] <http://www.unep.org/regionalseas/programmes/nonunep/redsea/default.asp> accessed 15/3/16.

In 1982 the Programme was underpinned by the *Red Sea/Gulf of Aden Convention*, which was the result of a Regional Intergovernmental Conference held in Jeddah during 13–15 February 1982.[107] In the same way as the *Arabian Gulf/Gulf of Oman Convention*, this responded to the LOSC,[108] with Article 123 calling for cooperation among coastal countries of an enclosed or a semi-enclosed sea, and for them to coordinate their efforts regarding rights and duties concerning marine environment protection and maintaining it from pollution. The geographical coverage of the *Red Sea/Gulf of Aden Convention* is stipulated as applying to 'the entire sea area, taking into account integrated ecosystems of the Red Sea, Gulf of Aqaba, Gulf of Suez, Suez Canal to its end on the Mediterranean, and the Gulf of Aden' (Article II). The Conference also resulted in two other legally binding instruments: the *Action Plan for the Conservation of the Marine Environment and Coastal Areas in the Red Sea and Gulf of Aden ('Action Plan')*;[109] and the *Protocol Concerning Regional Cooperation in Combating Pollution by Oil and Other Harmful Substances in Cases of Emergency ('Regional Cooperation Protocol')*.[110]

The aims of the PERSGA *Action Plan* are identical to the ROPME *Action Plan* outlined in an earlier section. It also consists of the same four main components, which are common to all of the Regional Seas Programmes. Gladstone et al. comment on the PERSGA arrangements as follows:

> The SAP [Strategic Action Plan] has identified a series of issues and the underlying causes need to be addressed by specific regional actions … The actions reflect the critical need for integrated coastal zone management and the need to integrate environmental planning, environmental assessment and review procedures into the development planning process. The actions address the strengthening of responsible regional and national institutions and the development of human capacity required for environmental management.[111]

---

[107] PERSGA, Final Act of the Jeddah Plenipotentiary Regional Conference on the Conservation of the Marine Environment and Coastal Areas in the Red Sea and Gulf of Aden (Jeddah, 13–15 February 1982).

[108] Above n 39.

[109] PERSGA, Action Plan for the Conservation of the Marine Environment and Coastal Areas in the Red Sea and Gulf of Aden (Jeddah, 14 February 1982) in force 20 August 1985, <http://www.persga.org/Documents/Doc_62_20090211114435.pdf> accessed 13/3/15.

[110] PERSGA, Protocol Concerning Regional Cooperation in Combating Pollution by Oil and Other Harmful Substances in Cases of Emergency (Jeddah, 14 February 1982) unreported, in force 20 August 1985.

[111] William Gladstone, N Tawfiq, D Nasr, I Andersen, C Cheung, H Drammeh, F Krupp and S Lintner, 'Sustainable Use of Renewable Resources and

In September 1995 PERSGA formulated three additional instruments: the *Protocol Concerning the Conservation of Biological Diversity and the Establishment of a Network of Protected Areas in the Red Sea and Gulf of Aden* (*'Biodiversity Protocol'*);[112] the *Protocol Concerning the Protection of the Marine Environment from Land-Based Activities in the Red Sea and Gulf of Aden* (*'Land-Based Activities Protocol'*);[113] and the *Protocol Concerning the Exchange of Personnel and Equipment in Case of Marine Emergency* (*'Exchange of Personnel and Equipment Protocol'*).[114] Memoranda of understanding were also prepared and signed with relevant regional and international organizations including ALECSO, ROPME, the Islamic Educational, Scientific and Cultural Organization (ISESCO), UNEP, and the IMO; a Regional *Action Plan* for contingency planning was prepared in cooperation with IMO and was subsequently endorsed by PERSGA member states.

Following the structure of the *Arabian Gulf/Gulf of Oman Convention* closely, PERSGA is established similarly to ROPME under Article XVI(1) of the *Red Sea/Gulf of Aden Convention*. In an almost identical way, it is also stipulated to consist of three parts, the latter two with slightly different names: a Council, with a representative of each Party; a General Secretariat; and a Committee for the Settlement of Disputes (Article XVI(2)). International legal personality is granted to PERSGA in accordance with Article XVI(3).[115] Further detail in relation to the operation of the Council, meetings and voting procedure is set out in Article XVII; and the duties and functions of the Council in Article

---

Conservation in the Red Sea and Gulf of Aden: Issues, Needs and Strategic Actions' (1999) 42 *Ocean and Coastal Management* 671, 690.

112 PERSGA, Protocol Concerning the Conservation of Biological Diversity and the Establishment of Network of Protected Areas in the Red Sea and Gulf of Aden (Jeddah, 12 December 2005) Ecolex B7 p. 982:13/B, not in force.

113 PERSGA, Protocol Concerning the Protection of the Marine Environment from Land-Based Activities in the Red Sea and Gulf of Aden (Jeddah, 25 September 2005) unreported, not in force.

114 PERSGA, Protocol Concerning the Exchange of Personnel and Equipment in Case of Marine Emergency (Jeddah, 1 July 2009) unreported, not in force.

115 On the significance of this, see Sébastien Duyck, *Legal Issues Related to Options for a Regional Seas-Type Arrangement for the Arctic Ocean*, Northern Institute for Environmental and Minority Law; Eric C Ip, 'The Power of International Legal Personality in Regional Integration' UNU-CRIS Working Papers W-2010/4 (United Nations University, Brugge, 2010) 1–22; and Simon Marsden, 'From the High North to the Roof of the World: Arctic Precedents for Third Pole Governance' (2016) 8 *The Yearbook of Polar Law* 56–75.

XVIII. In relation to the latter, these include keeping under review the implementation of the Convention, protocols and *Action Plan* (Article XVIII(b)); making recommendations in relation to additional protocols (Article XVIII(c)); adopting, reviewing or amending annexes to the Convention or protocols (Article XVIII(d)); adopting or concluding agreements with States or organizations with similar purposes or interests (Article XVIII(e)); reviewing or evaluating the state of the marine environment and coastal areas further to reports from Parties or international organizations (Article XVIII(f)); establishing subsidiary bodies (Article XVIII(g)); and considering NIRs from Parties (Article XVIII(h)).

3.2 Objectives, Principles and Procedures, including Implementation and Compliance

3.2.1 Convention

The objective of the *Red Sea/Gulf of Aden Convention* as expressed in the general obligations (Article III(1)) is 'the *conservation* of the Red Sea and Gulf of Aden environment including the prevention, abatement and combating of marine pollution' (emphasis added). Reflective of the passage of time and greater attention to environmental conservation, this differs from the wording of the *Arabian Gulf/Gulf of Oman Convention*, which makes no reference to 'conservation' in the comparable provision (Article III(a)). Despite this, the approach is however virtually identical, with national standards, laws and regulations required for implementation, as well as the harmonization of related national policies, and the appointment of a National Authority by each Party to coordinate such efforts (Article III(3)).

The recitals to the Convention furthermore draw attention to urban and rural development and land use and the need to 'preserve, as far as possible, marine resources and coastal amenities'; 'ensure that the processes of industrial development should not, in any way, cause damage to the marine environment, jeopardize its living resources or create hazards to human health'; recognize 'the need to develop an integrated management approach to the use of the marine environment and the coastal areas', and 'the need for a carefully planned research, monitoring and assessment programme'; and finally, 'the importance of cooperation and coordination on a regional basis', and to protect the marine environment 'for the benefit of all concerned, including future generations'. As identified by Khan and Al-Ajmi in relation to the

Arabian Gulf/Gulf of Oman Convention,[116] the Red Sea/Gulf of Aden are also semi-enclosed seas and collectively a pollutant trap. Gladstone et al. explain the geo-physical dynamics and comment on the challenges posed by this as follows:

> Semi-enclosed seas are globally unique ecosystems. They are also susceptible to disturbances from human activities, and the sharing of their coasts and catchments by several countries presents unique challenges to environmental management and conservation efforts. Limited exchange between semi-enclosed seas and the oceans increases the chances of environmental damage from pollution events and activities that compromise ecosystem integrity such as coastal and catchment habitat destruction, the loss of biodiversity, and the unsustainable use of resources.[117]

Other general obligations (again shared with the *Arabian Gulf/Gulf of Oman Convention*) require 'cooperation with the competent international, regional and subregional organization to establish and adopt regional standards [and] recommend practices and procedures' (Article III(4)) and ensure that implementation does not 'cause transformation of one type or form of pollution to another which could be more detrimental to the environment' (Article III(5)). The general obligations are supported by the same specific, one paragraph, obligations as the *Arabian Gulf/Gulf of Oman Convention*, in relation to the following: pollution from ships (Article IV), pollution caused by dumping from ships or aircraft (Article V), pollution from land-based sources (Article VI), pollution resulting from exploration and exploitation of the bed of the territorial sea, the continental shelf and the subsoil (Article VII),[118] and pollution from other human activities (Article VIII). The need for Parties 'to cooperate in the formulation and adoption of appropriate rules and procedures' in relation to liability and compensation is expressed in Article XIII of the Convention, 'bearing in mind applicable international rules and procedures' (Article XIII(1)), 'for damage resulting from violation of obligations under the present Convention and its protocols' (Article XIII(2)).

---

[116] Above n. 44.

[117] Above n. 111, 672.

[118] The same ROPME/MEMAC guidelines are applicable in relation to seismic operations, and the storage of chemicals. See above nn. 89 and 88 respectively.

Other procedural provisions in the *Red Sea/Gulf of Aden Convention* include requirements for cooperation in dealing with pollution emergencies (Article IX), scientific and technological cooperation (Article X),[119] and assessment and management of the environment (Article XI), and for providing technical and other assistance (Article XII). In relation to Article IX, Parties are required to ensure that adequate equipment and qualified personnel are available to deal with such emergencies (Article IX(1)); that a Party that becomes aware of such an emergency notify the Organization and any Contracting Party likely to be affected (Article IX(2)); and that Parties coordinate national plans in a way that allows for full cooperation (Article IX(3)); the *Regional Cooperation Protocol* and *Exchange of Personnel and Equipment Protocol* contain more detailed obligations in relation to these emergencies (see below).

In relation to Article XI, the intention is, unlike Article IX – which is triggered after damage has arisen – to take a preventive approach by giving 'due consideration to marine environmental effects when planning or executing projects' (Article XI(1)), and, in support of this obligation, to 'develop procedures for dissemination of information on the assessment of the[se] activities'. 'Due consideration' is a stronger form of wording than the comparative provision under the *Arabian Gulf/Gulf of Oman Convention* (Article XI(a)), which only requires Contracting States to 'endeavour to include an assessment'. The emphasis on the preventative principle is significant because of the difference between this subregion and others. Although writing some time ago and subject to development and other activity since, Gladstone et al. highlight this:

> In comparison with other regional seas degraded by pollution and requiring costly curative responses, the majority of the actions needed in the Red Sea and Gulf of Aden are preventive. This reflects the generally good quality of the marine and coastal environments and the lack of development in many parts of the region.[120]

In general, however, it is fair to say that both the regimes analysed in this chapter do not fully incorporate environmental principles in the regulatory scheme. While the Jeddah regime – as the more recent – makes

---

[119] Nadim et al. (above n. 26, 563), report that in June 2005, Iran's Department of Environment was placed in charge of administrating comprehensive oceanographic studies of the Persian Gulf and Oman Sea with the assistance of experts from ROPME. Saudi Arabia and Kuwait have established research centres to study and preserve the fragile habitat of the Gulfs region.

[120] Above n. 111, 690.

reference to 'Cross-cutting Principles and Issues' on its website,[121] there is nothing similar in the Kuwait regime, either in the agreements or related documents, or indeed on the website of that regime. These cross-cutting principles are: poverty alleviation; mainstreaming environmental considerations into development proposals; applying the precautionary principle and eco-system management approaches; good governance and public participation; and 'on-the-ground activities' (essentially implementation of policies and plans).

NIR and compliance control and dispute settlement obligations are found in Articles XXII and XXIII respectively. In relation to NIR, Parties 'shall submit to the General Secretariat reports on measures taken for the implementation of this Convention and its protocols, in such form and at such intervals as may be determined by the Council'. It should be noted that Article III(3) requires the establishment of 'national standards, laws and regulations' in order to discharge the general conservation obligation. As seen in relation to the example of Bahrain in connection with the establishment of national EIA law in the previous section on the *Arabian Gulf/Gulf of Oman Convention*, this is not always easy in practice. Regarding compliance control, Parties are required to 'cooperate in the development and implementation of procedures for the effective application of the Convention and its protocols, including detection of violations, using all appropriate and practicable measures of detection and environmental monitoring, including adequate procedures for reporting and accumulation of evidence'.

Finally, procedures for the settlement of disputes are found in Article XXIV. This differs from the section in the *Arabian Gulf/Gulf of Oman Convention* (Article XXV), which does not involve the Council in deliberations if any dispute cannot be settled amicably, but rather proceeds directly to the Judicial Commission if agreement cannot be reached. The next section considers provisions in the four related agreements: the *Regional Cooperation Protocol, Biodiversity Protocol, Land-Based Activities Protocol* and the *Exchange of Personnel and Equipment Protocol.*

3.2.2 Protocols

The experience gained from the implementation of the Arabian Gulf/Gulf of Oman regime has been invaluable in the development of the Red Sea/Gulf of Aden regime, particularly the development of protocols under the latter. The importance of developing capacity for strengthening

---

[121] See: <http://www.persga.org/inner.php?id=338> accessed 13/5/16.

institutions underlies these efforts, as noted.[122] The *Exchange of Personnel and Equipment Protocol* is the most recent addition to the regime and therefore responded to the absence of such an agreement in the Arabian Gulf/Gulf of Oman regime. This protocol is considered an important supplement that consolidates the Jeddah Convention and the activities of PERSGA; it also plays an important role in promoting regional cooperation to reduce the risk of pollution resulting from a maritime accident. It includes 11 articles and three annexes.[123] In connection with the first of the protocols, the *Regional Cooperation Protocol*, there is considerable duplication of the provisions of the related agreement with the same name under the *Arabian Gulf/Gulf of Oman Convention*. Other than as stated below, the provisions of this agreement are effectively the same as in its predecessor agreement, and most of the differences are of little consequence.

For example the word 'resources' has been added to the definition of 'Marine Emergency Contingency Plan' in Article 1(3), which doesn't change the meaning of the comparative provision in the earlier Protocol under the *Arabian Gulf/Gulf of Oman Convention*. Similarly, the example given for the definition of 'Related Interests' in Article 1(5)(c) no longer includes the expression 'and the well-being of the area concerned' immediately after 'health of the coastal population' which is present in the earlier Protocol, which is unlikely to change the meaning significantly. Similarly, the absence of the subparagraph (c) to Article VI of the former agreement 'Results of research and development referred to in paragraph (b) above' does not affect the meaning of the provision in this Protocol.

One notable absence from the more recent Protocol is however Appendix A, which is present in the earlier agreement: 'Guidelines for the report to be made pursuant to Article VII of the Protocol', where the report concerns marine emergencies; in the more recent agreement, it is up to the MEMAC to adopt these. One provision that is common to both Protocols, which is not defined but is clearly of some significance, is Article XI(5), where 'In cases of special emergencies, the Centre (MEMAC) may call for the mobilization of resources made available by the Contracting States to combat pollution by oil and other harmful substances'. The Gulf War would clearly have been a 'special emergency' on any reckoning, but this occurred before both Protocols were

---

[122] See Gladstone et al., above n. 111, 690.

[123] The more recent Protocol is available only in Arabic and is not discussed. See: 'Technical Cooperation Protocol', available at: <http://www.persga.org/inner.php?id=62> accessed 14/3/16.

adopted, so it is unclear what exactly this means and why it is present. What is significant, however, is the recognized need to prepare for contingencies, which is addressed by this protocol. Gerges comments on one feature in particular:

> The Regional Contingency Plan is intended to facilitate the exchange of information on the emergency cases between the states concerned, and the movement and use of equipment, including vessels, aircraft and oil spill combating equipment. It also sets up clear definition of command structure and liaison arrangements for joint response operations, etc.[124]

As to the second protocol, the *Land-Based Activities Protocol*, although the *Arabian Gulf/Gulf of Oman Convention* also has a protocol with a very similar name (*Land-Based Sources Protocol*), there are more significant differences between the two. Therefore the major provisions of the latter agreement are analysed here, with indications of differences where they are apparent. Unlike the *Land-Based Sources Protocol*, the objective of the *Land-Based Activities Protocol* is explicit in Article 1, which requires Parties to:

> … take all appropriate measures to protect the environment of the Red Sea and Gulf of Aden against pollution resulting from any land-based sources or activities and to reduce and/or eliminate such pollution to the maximum extent possible with priority given to the gradual elimination of toxic, persistent, and biologically accumulating inputs.

This reflects the knowledge gained and physical development since the earlier protocol, which was adopted in 1990 in comparison with 2005 for the later one. It is seen furthermore in the Recitals to the later protocol, which note the 'increased pressures on the environment resulting from human activities … mainly due to dredging and land filling connected with urban and industrial development in coastal areas in addition to the seasonal increase of population in certain areas of tourist significance' (Recital 2). It continues '… these problems are essentially due to discharge of sewage, untreated or partially treated industrial effluents, or effluents discharged in a manner unfavourable to the receiving environment …' (Recital 4).

The Recitals of this protocol also recognize other relevant international developments and provisions – in addition to the LOSC, which is also

---

[124] Makram A Gerges, 'The Red Sea and Gulf of Aden Action Plan – Facing the Challenges of an Ocean Gateway' (2002) 45 *Ocean and Coastal Management* 885, 892.

cited by the Recitals in the earlier protocol – in particular the GPA for the Protection of the Marine Environment for 1995. This is defined, alongside the Assessment Report of UNEP concerning assessment of pollution from land-based sources and activities, with the more modern formulation 'biologically accumulative, persistent toxic substances', and 'land-based activities' – 'any human land activity which results, directly or indirectly, in pollution of the marine environment and exposes living and non-living natural resources to destruction or threat' – in Article 2. The influence of the *Convention on Biological Diversity* (CBD)[125] – adopted in 1992 after the first Protocol – can also be seen in these definitions.

Article 4 of the Protocol indicates that the scope and application of the Protocol covers both 'discharges and emissions of pollutants and radioactive substances' and 'air-borne pollutants'. General obligations are set out in Article 5, which include pollution prevention and elimination further to the implementation of work plans based on source control; source control is detailed in Annex II (Article 5(1)). National and regional work programmes and plans are required (Article 5(2)); with priorities and implementation timetables approved by the Parties and reviewed by the PERSGA Council (Article 5(3)); the best available technologies and environmental practices including clean production technology taken into account (Article 5(4)); and preventive measures taken where possible to reduce pollution hazards (Article 5(5)). It is important to recognize, however, that in connection with prevention, this subregion has historically been both undeveloped and – in contrast to the states of the Arabian Gulf/Gulf of Oman subregion – relatively unsupported by member states. Capacity has therefore been a major issue. Gerges explains:

> The environmental problems in the Red Sea and Gulf of Aden Region differ somewhat from the other Regional Seas. The combination of severe underdevelopment of some states of the Region, coupled with extreme and rapid economic expansion in others, has created a peculiar setting in which the Region's environmental agencies have to deal with a wide range of existing and emerging environmental problems at all levels, local, national and regional. This constitutes a real challenge to the Region, particularly in view of the limited resources available and the insufficient technical and financial capacities of most countries to deal with such a complex situation in which environmental problems directly and indirectly interact with the socio-economic difficulties known to prevail in the Region.[126]

---

[125] Above n. 54.

[126] Ibid, 893. These socio-economic problems may be implicated in the emergence of the problem of Somali piracy considered in the literature. See:

Specific obligations follow in relation to the treatment and management of used water, in compliance with EIA and other matters (Article 6); the management of solid wastes to ensure their elimination, further to the UNEP GPA (Article 7); and sedimentation and dredging practices threatening the marine environment, also guided by the UNEP GPA (Article 8). Article 9 concerns the protection of habitats, in recognition of how 'development and urban expansion … increase the rate of negative changes'; and therefore requires Parties to apply Integrated Coastal Zone Management (ICZM) in the protection of habitats (Article 9(1)).[127] Although – like EIA – ICZM is not defined in the Protocol (but see Article 13 considered below) it is a well-used environmental management tool that is clearly understood.[128] The provision also requires Parties to provide technical and institutional support where needed (Article 9(2)), and encourage conservation of environmentally sensitive areas (Article 9(3)).

Provisions follow in relation to national legislation and regional guidelines for waste disposal (Article 10);[129] licensing and waste disposal regulations (Article 11); environmental monitoring and data management (Article 12); EIA (Article 13); scientific and technological cooperation (Article 14); scientific and technical assistance (Article 15); and financing. For example, Article 13 requires Parties to incorporate EIA 'as a priority' (Article 13(1)), develop guidelines and other procedures (Article 13(2)), and establish a 'Regional Chamber' at PERSGA for information exchange with project studies accessible to Parties to save money and avoid duplication (Article 13(3)). Areas for priority protection shall also be determined, 'such as cultural heritage sites, areas of coral reefs,

---

James Kraska, 'Coalition Strategy and the Pirates of the Gulf of Aden and the Red Sea' (2009) 28 *Comparative Strategy* 197, 208 who comments on the 'environmental hardship' referred to in the recent Nairobi Report as one factor contributing 'to the rise and continuance of piracy in Somalia'; and also Marie Christine Boilard, 'Improving Policy Responses to Piracy in the Red Sea and Gulf of Aden: What Role for India?' (2010) 34(4) *Strategic Analysis* 625–638, who highlights the role of illegal fishing activity in reducing the livelihoods of local people (at 627).

[127] In connection with damage to coral reefs, see: ROPME, *Guidelines for Compensation Following Damage to Coral Reefs by Ship or Boat Grounding*, Technical Series Number 15, February 2009.

[128] See Ali Pak and Farajzadeh Manuchehr, 'Iran's Integrated Coastal Management Plan: Persian Gulf, Oman Sea, and Southern Caspian Sea Coastlines' (2007) 50 *Ocean and Coastal Management* 754, for example.

[129] See above n. 87.

mangroves and seagrass beds which could be declared as marine protectorates and where establishment of projects would be prohibited' (Article 13(4)). The PERSGA Secretariat is obliged to prepare a list of specialists with expertise in ICZM to assist with the implementation of EIA (Article 13(5)). General guidelines have been produced,[130] specifically in relation to fishery and aquaculture projects which support implementation efforts.[131]

Final provisions concern transboundary pollution, and dispute settlement and compensation (Article 17); reporting (Article 18); the regional work programme (Article 19); and implementation (Article 20). Article 17 requires consultations to take place between Parties when pollution originates in the territory of one of them and affects another or others, with a view to an 'amicable settlement' being reached. NIRs are required to be submitted by Parties to the PERSGA Council every two years (Article 18). In accordance with Article 19, the Programme is 'designed to assess the impacts of pollutants on human health and ecosystems' and is 'the basis for regional cooperation'; the priorities to be taken into account are specified in Annex I. Compliance control is ultimately the task of the Council, which is given the authority to 'Follow up the Protocol's implementation and scrutinize the effectiveness of the approved programmes, measures and work plans' among other matters (Article 20(a)).

The third protocol under the *Red Sea/Gulf of Aden Convention* is the *Biodiversity Protocol.* Like the *Land-Based Activities Protocol*, this was also adopted in 2005. In addition to the concerns expressed about the continued pressures on the natural environment, and the reference to the CBD, and application of the precautionary and common but differentiated responsibility principles in the Recitals, Article 1 states two objectives of the Protocol:

1. To provide for the conservation, protection and restoration of the health and integrity of the ecosystems and biological diversity in the PERSGA region.

---

[130] ROPME, *Guidelines on Requirements for the Conduct of Environmental Impact Surveys and the Production of Environmental Impact Statements*, MEMAC, undated.

[131] Regional Organization for the Conservation of the Environment of the Red Sea and Gulf of Aden (PERSGA), *EIA and EMS Guidelines for Fishery/ Aquaculture Projects in the Red Sea and Gulf of Aden*, PERSGA Technical Series No 9, May 2004.

2. To safeguard the threatened species, the critical habitats, sites of particular importance, as well as representative types of coastal and marine ecosystems, their biodiversity and their sustainable use and management, to ensure long-term viability and diversity.

Following definition of terms and area of application in Articles 2 and 3, Article 4 expresses the general obligations of the Protocol. These require 'all appropriate measures' to: protect, conserve and manage biodiversity, especially threatened species (Article 4(1)), and in 'unique, highly sensitive or regional representative' areas (Article 4(2)); adopt strategies, plans and programmes (Article 4(3)); adopt appropriate planning, management and supervision including legislation and monitoring measures (Article 4(4));[132] adopt comprehensive EIAs (Article 4(5)) to evaluate mariculture operations and their consequences (Article 4(6));[133] control pollution with potential significant impacts (Article 4(6)); ensure tenure provides for intergenerational equity (Article 4(7)); and designate competent authorities (Article 4(8)). Species protection and conservation is the subject of Part 2 of the Protocol, while Part 3 concerns the protection of selected regionally representative marine and coastal areas.

In relation to Article 4(1), Article 5 establishes a List of Threatened Species, requiring the application of the IUCN criteria[134] to the national inventories to be prepared under Article 5(1), which is to constitute Annex I to the Protocol. Article 5(2) invites Parties to 'consider developing and applying common, preferably qualitative criteria to determine species which are threatened'; Article 5(3) requires Parties to account for conservation measures in relation to threatened species; and Article 5(4) suggests that monitoring take place periodically by thc Parties 'according to common criteria such as population size, extent of occurrence, number of mature individuals and estimated immediacy of extinction'. Article 6 requires Parties to provide lists of aquatic species of commercial and/or cultural value; these are to include population status and measures to

---

132 In relation to strategic planning, see: Regional Organization for the Conservation of the Environment of the Red Sea and Gulf of Aden, *The Red Sea and Gulf of Aden Regional Network of Marine Protected Areas – Regional Master Plan*, PERSGA Technical Report Series No. 1, PERSGA Jeddah 2002; and Regional Organization for the Conservation of the Environment of the Red Sea and Gulf of Aden, *Regional Action Plan for the Conservation of Coral Reefs in the Red Sea and Gulf of Aden*, PERSGA Technical Report Series No. 3, PERSGA Jeddah 2003.

133 See guidelines above nn. 130 and 131.

134 See 'The Red List' at: <http://www.iucnredlist.org/technical-documents/categories-and-criteria> accessed 12/5/16.

address population decline; they are to constitute Annex II to the Protocol. Article 7 concerns the maintenance of threatened species and species of economic and cultural importance, including their habitats (Article 7(2)); action should be at a regional level to address the migratory nature of many of them (Article 7(3)). Finally, regulation of alien species and genotypes is addressed by Article 8, requiring prevention, prohibition, monitoring and reporting.

In regard to Article 4(2), the establishment of a List of Protected Areas is required by Article 9 of the Protocol, which is referred to as the 'PERSGA List'. Among other things, this is to include sites important for conserving biological diversity in the region, ecosystems specific to the region, or habitats of threatened species (Article 9(2)). Article 10 outlines the procedure for inclusion of proposed protected areas in the PERSGA List (Article 10(4)), with proposals to include a statement justifying the importance of the area (Article 10(3)). Article 11 deals with the management of protected areas, to follow management plans and include maintenance of threatened species or those with economic or cultural importance; maintenance and conservation of special habitats; restoration of ecosystems and species populations; and involvement of communities. Article 12 encourages Parties to provide inventories of areas of special importance; and Article 13 the management of special habitats.

The establishment of Marine Protected Areas (MPAs) has been a significant part of the work undertaken in accordance with both the *Biodiversity Protocol* and CBD.[135] Designed to conserve representative examples of biodiversity or protect unique features of international significance, PERSGA has therefore had an important role in this process, which in turn has been supported by ROPME, despite the absence of a subsidiary biodiversity agreement under the *Arabian Gulf/ Gulf of Oman Convention*. Gladstone et al. examine the regional network of MPAs in the Red Sea/Gulf of Aden as a case study in site selection, the development of regional management, the potential benefits for sustainable resource use and conservation, and the potential constraints to effective implementation.[136] These were established 'to manage human uses, to support sustainable resource usage and economic development, and to conserve representative and significant biodiversity'.[137] Gladstone et al. find that despite 75 MPAs being established or recommended, 'few

135 See *Regional Master Plan*, above n. 132.

136 William Gladstone, Fareed Kruup and Mohammed Younis, 'Development and Management of a Network of Marine Protected Areas in the Red Sea and Gulf of Aden Region' (2003) 46 *Ocean and Coastal Management* 741, 743.

137 Ibid, 746.

are managed appropriately … [and] there are gaps within these MPAs in representation of regionally significant and representative habitats'.[138] Furthermore, despite the obligation in Article 4(4) of the Protocol to enact legislation, the authors find that (at the time of publication) only Egypt and Saudi Arabia had enacted framework laws to establish and manage protected areas, although Jordan, Sudan and Yemen had, in addition to Egypt and Saudi Arabia, also introduced specific laws for individual MPAs.[139]

Gladstone updated the findings of the above research five years later in an editorial for another journal. He found that there had been progress in implementation, with management plans initiated for MPAs established in Djibouti and Sudan, that training methods had been improved to aid assessment, and that two subregional Fisheries Research and Training Centres had been established in Aden and Jeddah.[140] Furthermore, he comments that there 'have been substantial gains in scientific knowledge', 'action plans [have been] developed to facilitate national implementation of the regional needs', 'implementation is occurring through integrated networks', and 'a regional database for key habitats and species [was] established in PERSGA'.[141] Of the weaknesses identified, however, 'Extensive coastal development and commercial benthic trawl fisheries mean there is an increasing need for monitoring', 'institutional and technical capacities for research, monitoring and stock assessments remain weak', a 'number of areas proposed … have not been officially declared and some … require site-specific management plans. The potential performance of some MPAs is compromised by a general lack of surveillance and enforcement, implementation and management plans, and management expertise'.[142]

Part 4 contains provisions common to protected areas, habitats and species; Part 5 contains other provisions including access to genetic resources, access to and transfer of technology and benefit distribution, and for scientific and technical cooperation, information exchange and environmental awareness, and – very briefly – NIRs (Article 23). Regarding Part 4, in contrast to the *Land-Based Activities Protocol*, Article 14 makes more specific reference to ICZM, with which it is

---

[138] Ibid, 746.

[139] Ibid, 756.

[140] William Gladstone, 'Editorial: Towards Conservation of a Globally Significant Ecosystem: The Red Sea and Gulf of Aden' (2008) 18 *Aquatic Conservation: Marine and Freshwater Ecosystems* 1, 3.

[141] Ibid, 3–4.

[142] Ibid, 4.

concerned, suggesting Parties 'consider adopting and applying principles of integrated coastal areas management, to provide an umbrella mechanism for biodiversity conservation ...'.

The remainder of the provision explains the requirements for these principles: early consideration of environmental factors, limiting resource use, understanding carrying capacity, avoidance of conflict, and the establishment and management of protected areas. EIA is addressed in Article 15, which is to include consideration of transboundary effects. The restoration of ecosystems and species populations is dealt with in Article 16, which requires Parties to take into account 'internationally or regionally recognized guidelines', consideration of artificial habitats, and monitoring of effectiveness. Article 17 is concerned with guidelines and common criteria generally, requiring adoption of IUCN criteria and other matters in relation to the Annexes of the Protocol. The overall assessment of the operation of the Protocol is expressed by Gladstone as follows:

> Continued progress requires some fundamental developments in a number of key areas: poorly developed environmental and fisheries legislation in some countries, lack of funding (for research, management, monitoring, surveillance and enforcement), lack of political will to implement management, and lack of expertise and experience in marine conservation. There is minimal enforcement of regulations and penalties for infringements are too low to act as an effective deterrent and to encourage compliance. Coordination between ministries responsible for fisheries, conservation and the environment is weak or inadequate and the agencies carry out their activities independently.[143]

4. CHINA'S RELATIONSHIP WITH SOUTHWEST ASIA

Unlike the regimes examined in Chapters 4 and 5 (for the Mekong[144] and Tumen[145] river basins in particular, and also with ASEAN[146]), China does not have a formal role in relation to either of the treaty regimes analysed in this or the next chapter concerning central Asia. However, the relationship between China and certain states of the southwest subregion

---

143 Ibid, 4–5.

144 China is a dialogue partner of the Mekong River Commission. See: <http://www.mrcmekong.org/about-mrc/upstream-partners/> accessed 18/5/16.

145 China is a member of the Tumen Programme. See: <http://www.cn.undp.org/content/china/en/home/operations/projects/poverty_reduction/northeast-asia-regional-cooperation-programme.html> accessed 14/11/16.

146 China is a dialogue partner of ASEAN. See: <http://www.asean.org/asean/external-relations/china/> accessed 18/5/16.

is growing, as China's need for oil is met by Saudi Arabia in particular;[147] this is a relationship of benefit also to Saudi Arabia as demand for oil globally dramatically declines, and its need to find other markets for oil grows. Saudi Arabia, bordering both the Arabian Gulf and Red Sea, is the only state Party to both of the two treaties analysed in this chapter, and is strategically located between them. With a population of around 32 million people and known as the birthplace of Islam, it is therefore arguably the most influential state in the region.

Iran's significance, however – in connection with both the *Arabian Gulf/Gulf of Oman Convention* in this chapter, and the *Caspian Sea Convention* in the next, is in some ways comparable. The relationship between China and Iran is explained in more detail in Chapter 3. This has developed in particular as part of, and as the result of, the end of Iran's isolation following the termination of its nuclear programme;[148] China's desire for oil and gas is also therefore directed at Iran, which in the case of oil production especially provides inevitable competition with other suppliers, since Iran refused to freeze oil production alongside other member states of the Organization of the Petroleum Exporting Countries (OPEC) group, arguing the need to regain market share after its isolation.[149]

The state visit by President Xi of China in January 2016 resulted in an expansion of bilateral ties with both Saudi Arabia and Iran.[150] In relation to Saudi Arabia, a comprehensive strategic partnership and boost to industrial capacity cooperation was a key objective. The two state oil producers, Saudi Aramco and Sinopec signed a framework agreement estimated to be worth between US$1 billion and US$1.5 billion as a clear demonstration of these ties.[151] A report of the China visit provides

---

[147] Pinghui Zhuang, 'Key Facts Behind China's Warming Ties with Saudi Arabia, Iran and Egypt as Xi Jinping Signs Mega Oil Deals During his Middle East Tour', *South China Morning Post*, 20 January 2016.

[148] For discussion in various areas, see: <http://www.nytimes.com/topic/subject/irans-nuclear-program> accessed 13/5/16. For legal background, see: Yaël Ronen, *The Iran Nuclear Issue* (Hart, Oxford, 2010).

[149] Graeme Wearden, 'Iran Urges Other Oil Producers to Freeze Output after Doha Debacle – As it Happened', *The Guardian*, 19 April 2016.

[150] In connection with the strategic relationship between Iran and Saudi Arabia – and the position of the other states in the subregion – see Shafeeq Ghabra, 'Between Iran and the Gulf: An Environment in Transition', *The Washington Institute for Near East Policy*, 16 May 2016. See also Thom Poole, 'Iran and Saudi Arabia's Great Rivalry Explained', *BBC News*, 4 January 2016.

[151] Above n. 147.

background to the developing relationship and notes the following outcomes, which have extended to security cooperation:

> The kingdom, once a staunch anti-communist nation and close ally of the United States, established diplomatic relations with China in 1990. Former Chinese president Jiang Zemin made the first state visit to Riyadh in 1999. Relations between the two nations have developed slowly, but many Saudi officials and businesses have already said there should be broader ties with China. Saudi Arabia is the biggest supplier of crude oil to China. The International Monetary Fund said that as of 2013, trade between the two nations has increased from US$1.28 billion in 1990 to US$74 billion in 2012. On the military front, China provided the kingdom with up to 60 CSS-2 intermediate-range ballistic missiles in 1988 – two years before they established diplomatic ties. Saudi Arabia also operates a medium-range DF-21 ballistic missile system, which has upset the US.[152]

The drastically reduced global demand for oil has also meant that Saudi Arabia has had to make cuts in various aspects of the economy,[153] together with the need to diversify into other non-oil related areas, such as coastal development.[154] This increased level of development is clearly of considerable concern for the effective functioning of both the *Arabian Gulf/Gulf of Oman* and *Red Sea/Gulf of Aden* regimes. One project recently announced is for the Red Sea Bridge, which is a transboundary proposal with its north African neighbour Egypt.[155] This demonstrates the intercontinental significance of the *Red Sea/Gulf of Aden* regime, which is also true of the first regime analysed in Chapter 3 for the Caspian Sea.

5. CONCLUSIONS

The geographical spread of the two regimes analysed in this chapter is impressive. They extend from the far southwest corner of Asia (including states on the African continent), to the centre of Asia to the northwest in

---

152 Above n. 147. In connection with terrestrial development proposals, Wong comments: 'Beijing is also in talks with more than 30 countries including … Saudi Arabia and Iran about high-speed rail projects'. See Peter Wong, 'The Dealmaker', *South China Morning Post*, 30 May 2016.

153 Robin Pagnamenta, 'Rise of Saudi Crown Prince', *The Times*, 23 April 2016.

154 Adam Taylor, 'Saudi Arabia Wants You to Go on Holiday There. Here's Why', *The Independent*, 27 April 2016.

155 Author unknown, 'Saudi Arabia and Egypt Announce Red Sea Bridge', *BBC News*, 8 April 2016.

the Caucasus bordering the European continent. It is noteworthy that their greatest concern is for the protection of coastal seas, and while operating autonomously, they are coordinated under the UNEP Regional Seas Programme. This Programme also includes the South Asian Seas[156] and East Asian Seas,[157] which are considered briefly in Chapter 5. The reason for this is that neither of these latter arrangements is underpinned by an international agreement of any kind, which differs from the focus of this chapter since both of the regimes analysed above are comprised of a regional treaty and associated agreements. The focus of the next chapter, Chapter 3, is however similar to the arrangements in Chapter 2, as the *Caspian Sea Convention*[158] both falls within the UNEP Regional Seas Programme, and it also consists of both a treaty and related protocols.

In connection with the *Arabian Gulf/Gulf of Oman* and the *Red Sea/Gulf of Aden* treaties, and as partly indicated by their full convention titles, the first is concerned with the protection of the Arabian Gulf and Gulf of Oman, and the second the Red Sea and the Gulf of Aden. Comparative analysis of both the regimes indicates almost identical provisions, with differences highlighted as appropriate. It was also noted that – as a result of the fairly early time at which particularly the *Arabian Gulf/Gulf of Oman Convention* was adopted – environmental principles guiding these provisions are either absent or largely non-explicit. This is in contrast to the later *Red Sea/Gulf of Aden Convention*, which makes reference to environmental principles in various places. In relation to both regimes, despite their existence, the opportunity for implementation and compliance measures to be followed or further developed over this time appears not to have been taken; information about practice with the reporting and compliance controls is therefore absent, and hence evaluation of effectiveness is difficult to achieve. Nonetheless, academic commentaries provide some indication of partial successes (and failures), as noted.

In addition to the environmental issues commented upon, security concerns apply to both of the subregions, for example in the Red Sea/Gulf of Aden, maritime piracy has been a major issue of concern, as

---

156 <http://www.unep.org/regionalseas/programmes/nonunep/southasian/default.asp> accessed 19/3/15.

157 <http://www.unep.org/regionalseas/programmes/unpro/eastasian/default.asp> accessed 19/3/15.

158 Framework Convention for the Protection of the Marine Environment of the Caspian Sea (Tehran, 4 November 2003), 44 ILM 1 (2005), in force 12 August 2006.

shown. As also indicated, there are clearly links between security and environmental issues where livelihoods have been impacted upon by development. For example, changes to or the destruction of traditional fishing practices has led to disengagement from society and involvement in activity such as piracy. In more recent times, disaffection has also resulted in involvement in terrorist activity, in particular support for Islamic State.[159] As will be seen in Chapter 3, security issues, both resource-focused and political in nature, are of key relevance to the arrangements for the Caspian Sea especially, and this has many other complexities, notably a Soviet past; yet members of the arrangements there are actively courted by many other states, in particular China, again because of the vast natural resources of the Sea, which has huge implications for the environmental protection objectives of the regime that is applicable to it.

---

[159] There is considerable comment on this in the online media, in response to fast-moving events. These issues are beyond the scope of this book, but the relationship between conflict and the environment is cause for a note at the least. As an example, see: <http://www.hudson.org/research/12291-saudi-arabia-s-islamic-alliance-major-challenge-for-al-baghdadi-s-islamic-state-or-potential-opportunity> accessed 13/5/16.

3. Connecting central Asia with southwest, north and Eurasia: the Caspian Sea, Aral Sea and sustainable development regimes

1. INTRODUCTION

From the coastal seas and gulfs of southwest Asia in Chapter 2, which connect the African and Asian continents, this chapter considers the arrangements primarily for the inland seas in central Asia, notably the Caspian and Aral Seas. In this part of Asia, there is a connection with Europe, given the role of Azerbaijan and Russia in the first regime, the *Framework Convention for the Protection of the Marine Environment of the Caspian Sea* (*Caspian Sea Convention*);[1] this regime also connects southwest and central Asia, with Iran – a Party to the *Arabian Gulf/Gulf of Oman Convention*[2] analysed in Chapter 2 – also a Party to the *Caspian Sea Convention.*[3] Russia and Iran, like Saudi Arabia – which is a Party to both the *Arabian Gulf/Gulf of Oman Convention* and *Red Sea/Gulf of Aden Convention*[4] – are not only the most economically powerful states in the Caspian Sea region, but are also geographically located in a place of major strategic importance. While the focus of this chapter is upon the

[1] Framework Convention for the Protection of the Marine Environment of the Caspian Sea (Tehran, 4 November 2003), 44 ILM 1 (2005), in force 12 August 2006.

[2] Regional Convention for Cooperation on the Protection of the Marine Environment from Pollution (Kuwait, 24 April 1978), in force on 1 July 1979 (the 'Arabian Gulf/Gulf of Oman Convention').

[3] For examples in connection with environmental management, see Ali Pak and Farajzadeh Manuchehr, 'Iran's Integrated Coastal Management Plan: Persian Gulf, Oman Sea, and *Southern* Caspian Sea Coastlines' (2007) 50 *Ocean and Coastal Management* 754–773.

[4] Regional Convention for the Conservation of the Red Sea and Gulf of Aden (Jeddah, 14 February 1982), ECOLEX B7 p. 982:13, in force 20 August 1985 (the 'Red Sea/Gulf of Aden Convention').

central Asian states because two of the three regimes are based in this subregion, Russia and Iran's significance – as with that of Saudi Arabia in the previous chapter – is without question when it comes to the ability to influence outcomes.[5]

The second regime is made up of the water resource arrangements for the Aral Sea, which are primarily and collectively constituted by: the *Agreement on Cooperation in the Field of Joint Management and Conservation of Interstate Water Resources* (*Interstate Water Resources Agreement*),[6] the *Agreement about the Status of the International Fund for Saving the Aral Sea* (*Saving the Aral Sea Agreement*)[7] and the *Statute of the Interstate Commission for Water Coordination of Central Asia* (*Interstate Commission Agreement*).[8] There are some similarities between these arrangements and those in south Asia that will be outlined briefly in Chapter 5. The third regime is the inactive *Framework Convention on Environmental Protection for Sustainable Development in Central Asia* (*Sustainable Development Convention*).[9]

The *Sustainable Development Convention* is one of a few of the regimes analysed in this book that is not in force (the other notable example is the *ASEAN Conservation Convention*);[10] it is particularly significant because unlike any of the regimes in this book it is a comprehensive treaty covering the full range of environmental issues. The control of the negative effects of natural resource exploitation that

---

[5] Although Iran has now been brought in from the cold after settlement of the nuclear proliferation issue in 2015, Kandiyoti, in discussing the role of the USA in the geopolitics of the area, comments that 'more carefully calibrated US policies could have made a profound difference on the regional chessboard by engaging with Iran, the idle hinge'. See Rafael Kandiyoti, 'What Price Access to the Open Seas? The Geopolitics of Oil and Gas Transmission from the Trans-Caspian Republics' (2008) 27(1) *Central Asian Survey* 75, 90.

[6] Agreement on Cooperation in the Field of Joint Management and Conservation of Interstate Water Resources (Almaty, 18 February 1992), accessed 20 November 2014 at <http://www.caee.utexas.edu/prof/mckinney/papers/aral/agreements/ICWC-Feb18-1992.pdf> accessed 27/1/16.

[7] Agreement about the Status of the International Fund for Saving the Aral Sea (Ashgabat, 9 April 1999), unreported.

[8] Statute of the Interstate Commission for Water Coordination of Central Asia (Almaty, 18 September 2008), unreported.

[9] Framework Convention on Environmental Protection for Sustainable Development in Central Asia (Ashgabat, 22 November 2006) TRE-143806, not in force, <http://www.ecolex.org/server2.php/libcat/docs/TRE/Full/En/TRE-143806.pdf> accessed 20 November 2014.

[10] Agreement on the Conservation of Nature and Natural Resources (Kuala Lumpur, 9 July 1985), not in force.

are part of the regulatory scheme, however, inform many of the environmental concerns (in the Caspian Sea region especially even if not the water use for agricultural purposes in the Aral Sea). Nonetheless, in relation to the analysis of each of the regimes, again subsections will outline the geographical context, historical background and institutional arrangements; objectives, principles and an overview of procedures, including matters of implementation and compliance; and, where possible, evidence of practice for the regimes in force. Practice will include examples of China's more active role in this subregion compared to southwest Asia, before conclusions are drawn.

1.1 Defining Central Asia

Under the UN Geoscheme, central Asia is clearly demarcated as comprising the five states of the subregion: Kazakhstan, Kyrgyzstan, Tajikistan, Turkmenistan and Uzbekistan. Although perhaps more geographically accurate (at least if the Russian Far East is left out of the equation)[11] than the term 'middle' East is for southwest Asia, its contours have nonetheless been subject to debate, as considered below. Kazakhstan has a population of 17 million, Kyrgyzstan 6 million, Tajikistan 8 million, Turkmenistan 5 million and Uzbekistan 30 million. Due to energy exports, Kazakstan is the wealthiest and Tajikistan the poorest. These dynamics, together with the 'upstream' or 'downstream' status of the country in question, influence the success or failure of any regime, as will be explored below. In relation to energy-poor Tajikistan for example, development of hydro-electric power is a key part of its policy future,[12] but this has implications for downstream states; similar issues affect the other mountainous state in central Asia, Kyrgyzstan.[13] As noted below and in Chapter 5, exploitation of water resources in these mountainous states also has implications for the protection of the Third Pole, which includes this mountainous terrain. Regulatory solutions have been found to depend not just upon the success or failure of national and subregional

---

[11] A brief look at a map shows that if Asia stretches from the Gulf of Aden in the southwest to the Russian northeast, then parts of China are the most 'central' in Asia or in the 'middle' of the 'East'.

[12] William C Rowe, 'Agrarian Adaptations in Tajikistan: Land Reform, Water and Law' (2010) 29(2) *Central Asian Survey* 189–204.

[13] Jeanne Féaux de la Croix, 'Moving Metaphors We Live By: Water and Flow in the Social Sciences and Around Hydroelectric Dams in Kyrgyzstan' (2011) 30(3–4) *Central Asian Survey* 487–502.

efforts, but upon the need to draw on resources from civil society and local communities.[14]

The subregion came into independent political and legal existence after the breakup of the Soviet Union in 1991,[15] and since independence it has had to establish relations with other states and organizations in a globalized world, as Russia was unable to assert its previously dominant position.[16] The constitutional system of each of the newly independent states of central Asia is an important part of the implementation of international law.[17] As with many of the states in this book, these are monist and do not strictly require the enactment of national law to ensure its implementation.[18] However, as seen in relation to the UNECE treaties considered in earlier work,[19] for treaty implementation to be effective, this is invariably needed nonetheless.

---

[14] As indicated by Féaux de la Croix's paper, ibid, useful background and insights can be drawn from social science scholarship more broadly. See further: Charles Buxton, 'NGO Networks in Central Asia and Global Civil Society: Potentials and Limitations' (2009) 28(1) *Central Asian Survey* 43–58; and Joe Hill, 'The Role of Authority in the Collective Management of Hill Irrigation Systems in the Alai (Kyrgyzstan) and Pamir (Tajikistan)' (2013) 33(3) *Mountain Research and Development* 294–304.

[15] Mirzohid Rahimov, 'From Soviet Republics to Independent Countries: Challenges of Transition in Central Asia', and Richard Pomfret, 'Central Asia since the Dissolution of the Soviet Union: Economic Reforms and their Impact on State-Society Relations', both in MP Amineh (ed), *The Greater Middle East in Global Politics: Social Science Perspectives on the Changing Geography of the World Politics* (Brill, Leiden, 2007) 291–332.

[16] See SJ Lloyd, 'Landlocked Central Asia' (1997) 2(1) *Geopolitics* 97; Emilian Kavalski, 'Uncovering the "New" Central Asia: The Dynamics of External Agency in a Turbulent Region', in Emilian Kavalski (ed), *The New Central Asia: The Regional Impact of International Actors* (World Scientific, Singapore, 2010) 6; and H Malik (ed), *Central Asia: Its Strategic Importance and Future Prospects* (St Martin's Press, New York, 1994).

[17] Scott Newton, *The Constitutional Systems of the Independent Central Asian States: A Contextual Analysis* (Hart Publishing, Oxford, 2016).

[18] Generally speaking, the majority of states considered in this book do not follow the common law tradition. This is true also of those which practise Islamic Law, which make up a large proportion of states not only in southwest Asia, but also in southeast Asia, such as Indonesia. For comprehensive information, see H Patrick Glenn, *Legal Traditions of the World: Sustainable Diversity in Law* (5th edn, Oxford University Press, Oxford, 2014); see also A Black, H Esmaeili and N Hosen, *Modern Perspectives on Islamic Law* (Edward Elgar Publishing, Cheltenham, 2013).

[19] Simon Marsden and Elizabeth Brandon, *Transboundary Environmental Governance in Asia: Practice and Prospects with the UNECE Environmental*

Qoraboyev considers the naming of central Asia as part of the 'world of regions'.[20] The author compares the term with two others, 'post-Soviet space', and 'central Eurasia', which reflect historical and geographical realities. The first was coined to refer to 12 of the 15 former Soviet republics – excluding the Baltic States, which obtained statehood in 1991 – and was a preferred alternative to the reference to the Commonwealth of Independent States (CIS) because of the view that the CIS was a failed organization. Central Eurasia in contrast has historically been used to describe territory ranging from Russia to Mongolia, Xinjiang and Tibet, as well as the five central Asian states and those in the south Caucasus.[21]

However, the fact that Chinese territory is included in any notion of Central Eurasia as well as Russia means it is not likely to be accepted; this is because of a Chinese preference for bilateralism in its foreign relations and due to the overwhelming focus on territorial integrity, matters which will be analysed in sections at the end of each chapter and in Chapter 5. Unless a distinction is drawn between the geographical and political theories of the idea of a Central Eurasia therefore, it will likely remain limited to use in the scholarly literature only; in particular, by authors looking at geopolitical actions by the significant actors in the area who vie for dominance, especially Russia, China and the USA,[22] in a process known as the 'Great Game'.[23] In contrast, Qoraboyev notes that

---

Agreements (Edward Elgar Publishing, Cheltenham, 2015). The influence of the UNECE in central Asia continues to grow with the organization of many capacity-building events, particularly in relation to water. See for example: United Nations Economic Commission for Europe, Chair's summary of the meeting 'Further development of regional cooperation among the countries of Central Asia in the management of information on the environment and water resources', Almaty, 9–10 June 2014, <http://www.unece.org/env/water/information_meeting.html#/> accessed 29/3/16. One of five recommendations of this event was the strengthening of legal and institutional frameworks, particularly in connection with the Aral Sea.

[20] Ikboljon Qoraboyev, 'Around the Names of Regions: The Case of Central Asia', UNU-CRIS Working Papers, W-2010/5 (United Nations University, 2010) 1, 4.

[21] See Murat Arsel and Max Spoor, *Water, Environmental Security and Sustainable Development: Conflict and Cooperation in Central Eurasia* (Routledge, London, 2010).

[22] See Matteo Fumagalli, 'The United States and Central Asia', in E Kavalski, *The New Central Asia: The Regional Impact of International Actors* (World Scientific, Singapore, 2010) 177–190.

[23] Ibid, 10. See also François Godement, 'Introduction', in European Council on Foreign Relations, *The New Great Game in Central Asia* (China Analysis/European Council on Foreign Relations/Asia Centre, 2011).

the 'central Asia' term was coined by the five relevant states themselves, after a meeting of their leaders in Tashkent in January 1993:

> Since then, a lively discourse around the 'Central Asia' region has emerged. This discourse consists of a set of ideas, representations and arguments aimed at both constructing and deconstructing Central Asia as a distinct region. The structure of this discourse can be summarized in a three-step dialectical outline. The first two blocks conceptualize Central Asia as one region in need of regional integration and a framework of cooperation, while the third set of arguments question the regionness of Central Asia and the viability of regional integration.[24]

The question of regional integration is especially pertinent to central Asia, and has implications for the success or failure of the regimes considered in the following sections. This is particularly true of the diffuse arrangements for the Aral Sea and the failure to advance the *Sustainable Development Convention*, which appears hostage to the collective political will of the five states. Efforts at regional integration began with a common economic zone established between the two largest states (Kazakhstan and Uzbekistan) in 1994, with Kyrgyzstan subsequently joining, and a decision being made to establish an Economic Union between them. After Tajikistan joined, however, a decision was taken to reduce competences, and the group was renamed the Central Asian Economic Community, albeit briefly before it was again renamed the Central Asian Cooperation Organization (CACO), with the idea of economic union shelved. Qoraboyev comments:

> The only dark spot in this image of Central Asia as a region with its own regional structures was the absence of Turkmenistan from the group since its beginning. In their devotion to establish a foreign policy based on positive neutrality, Turkmenistan's leaders always chose to stay away from different regional and supra-regional associations.[25]

Other setbacks followed, notably Russian involvement in the CACO and Uzbekistan joining a competing group, the Eurasian Economic Community (EAEC) – based within the CIS – in 2005. These two groups were subsequently merged, although this did little to dispel the view that the subregion lacked cohesiveness and had yet to emerge from Russian

---

[24] Qoraboyev, above n. 20, 12. For more detail, see the discussion that follows on central Asia as 'one region, regional integration and cooperation' and the unsuccessful attempts towards institutionalizing central Asia.

[25] Qoraboyev, above n. 20, 14.

hegemony.[26] The situation is extensively outlined by Qoraboyev, who highlights that one of the main issues has been the rivalry between the two largest states, Kazakhstan and Uzbekistan.[27] Furthermore, the strategic location of the subregion has been the source of ongoing tension between the most powerful external players, Russia and China, with the latter very successful in filling the power vacuum since the breakup of the Soviet Union, which will be considered at the end of this chapter. The following is a useful summary with reference to both energy and military security in the subregion, which makes reference to the role of the North Atlantic Treaty Organization (NATO):

> In world politics, Central Asia is identified as a region of geostrategic and geopolitical importance. It is identified as a strategic backyard both by Russia and China. At the same time it is located in the proximity of Afghanistan … and Iran, main spots of strategic instability and uncertainty in Eurasia. This element creates interest for Central Asia in countries involved in military operations in Afghanistan and in negotiations dealing with Iran. The US and the other NATO countries are the most interested. Moreover Central Asia is becoming increasingly viewed as an important component of the global energetic [sic] system. Central Asian resources in gas and oil, mainly situated in the Caspian Basin, were defined as second only to Persian Gulf resources.[28]

---

[26] See Marlène Laruelle, 'Russia and Central Asia', in E Kavalski, *The New Central Asia: The Regional Impact of International Actors* (World Scientific, Singapore, 2010); Ben N Dunlap, 'Divide and Conquer? The Russian Plan for Ownership of the Caspian Sea' (2004) 27(1) *Boston College International and Comparative Law Review* 115–130; this considers general as well as specifically legal issues relating to the LOSC. Note also Andrei Kazantsev, 'Russian Policy in Central Asia and Caspian Sea Region' (2008) 20(6) *Europe-Asia Studies* 1073–1088; this discusses the limits of cooperation between China and Russia at 6–7. And Evgeny Vinokurov and Alexander Libman, *Eurasian Integration Challenges of Transcontinental Regionalism* (Palgrave Macmillan, Houndmills, 2012), particularly 'Central Asia at the Crossroads: A Laboratory of Eurasian Integration', where the fluid dynamics between Russia and China are also discussed (at 214–222). Finally, see the comments of Russian President Putin in connection with integration: Vladimir Putin, 'Family of Nations: The Key to Progress and Development in the Asia-Pacific Region Lies in Ever-Closer Cooperation', *South China Morning Post*, 17 November 2015.

[27] Qoraboyev, above n. 20, 19.

[28] Qoraboyev, above n. 20, 19. In relation to the energy resources, see: US Energy Information Administration, *Caspian Sea Region* available at: <https://www.eia.gov/beta/international/regions-topics.cfm?RegionTopicID=CSR> accessed 17/3/16.

Turning to the environmental issues of the subregion,[29] oil and gas exploration and associated energy-related development is of primary importance;[30] given the participation also by southwest, north and Eurasian states (Iran, Russia and Azerbaijan), this is also the case under the *Arabian Gulf and Red Sea/Gulf of Aden* regimes. Niazi highlights their significance:

> Although Central Asia is landlocked, the tremendous untapped hydrocarbon wealth of the Caspian Sea region makes it the world's envy ... Geographically, Turkmenistan, Uzbekistan, Tajikistan and Kazakhstan make up the eastern side of the Caspian Sea Basin, beneath which lie oil reserves that rival those of Saudi Arabia and the world's richest reserves of natural gas.[31]

Central Asia is also subject to significant risk factors needing regulation if environmental impacts, and conflict over these resources, are to be avoided.[32] Vinogradov outlines an incident off the coast of Azerbaijan as an example, where a blowout in a gas-injection well on BP's Central Azeri platform resulted in the emergency evacuation of over 200 workers and shut down large parts of the offshore production in the Caspian's Azeri-Chirag-Guneshli (ACG) field. It also led to a significant decline in production, reducing Azerbaijan government revenues by up to $50 million per day. It was noted, however, that the financial loss was small in comparison to the $40 billion or more in cleanup costs and liabilities that British Petroleum (BP) faced over the Gulf of Mexico disaster; had it been a reservoir leak however, the damage would have been potentially non-reparable.[33] Vinogradov continues:

---

[29] For context, see: 'Transborder Ecological Issues on the Continent', in Alexander Libman and Evgeny Vinokurov, *Challenges of Transcontinental Regionalism* (Palgrave Macmillan, Houndmills, 2012) 141–150.

[30] Yesim Reel, 'Energy', in EAH Gencer and C Gerni (eds), *Central Asia Economies in Transition* (Cambridge Scholars, Newcastle, 2012); in relation to the Kashagan oil project, the subregion's most important, see Alexander Carrera, 'Kashagan: A Look at Caspian Sea Oil and Natural Gas Production', 23 September 2013; other information available at: <http://www.offshore-technology.com/projects/kashagan/> accessed 18/3/16.

[31] Tarique Niazi, 'The Ecology of Strategic Interests: China's Quest for Energy Security from the Indian Ocean and the South China Sea to the Caspian Sea Basin' (2006) 4(4) *China and Eurasian Forum Quarterly* 97, 110.

[32] Rob Johnson, *Oil, Islam and Conflict: Central Asia since 1945* (Reaktion Books, London, 2007).

[33] Sergei Vinogradov, 'The Impact of the Deepwater Horizon: The Evolving International Legal Regime for Offshore Accidental Pollution Prevention, Preparedness, and Response' (2013) 44(4) *Ocean Development and International Law*

> Given that the Caspian Sea is an enclosed body of water of relatively small size with an exceptionally valuable and vulnerable ecosystem, any large-scale spill could cause catastrophic irreversible damage to its marine environment, its living resources, and the economic interests of littoral states. The legal frameworks, international and domestic, in the Caspian Sea and other like regions are far from adequate to deal with oil pollution emergency situations. The very possibility of serious transboundary environmental impact in maritime areas criss-crossed by international boundaries calls for a thorough assessment of the applicable international law and its relevance to pollution caused by offshore accidents, especially in the context of emergency response and liability for pollution damage.[34]

The other major environmental issue for central Asia is water resource management in the Aral Sea Basin, where the upstream states Tajikistan and Kyrgyzstan have energy supply needs which are being met by hydro-electricity, and the downstream states of Kazakhstan, Uzbekistan and Turkmenistan have irrigation needs which are subject to the water supplied to them. Tajikistan and Kyrgyzstan therefore produce about 77 per cent of the inflow to the Basin, supplemented by about 10 per cent from Afghanistan, which is however not part of the regulatory arrangements in place. The development of hydro-electricity by the upstream states on one hand, and agricultural expansion of the downstream states on the other, has placed huge pressures on the Aral Sea which has diminished in size considerably. In relation to the first for example, the development of the Rogun Dam in Tajikistan is of considerable concern to Uzbekistan. As Putz comments:

> The dam project features prominently in the tense Uzbek-Tajik relationship. If built as planned, the dam will be the tallest in the world at 335 meters (1,099 feet) and have a capacity of 3200 MW. Situated on the Vakhsh River, a tributary of the Amu Darya, the dam will most certainly impact the volume of water flowing downstream.[35]

Furthermore, as McKinney states: 'inefficient irrigation systems and mismanagement of irrigation water diversions have resulted in elevated water and soil salinity levels, widespread environmental degradation, and

---

335, 339–340. See generally also: Hossein Esmaeili, *The Legal Regime of Offshore Oil Rigs in International Law* (Ashgate Dartmouth, 2001).

[34] Ibid, 362.

[35] Catherine Putz, 'Uzbekistan Still Hates the Rogun Dam Project', *The Diplomat*, 4 August 2015.

diminished agricultural productivity'.[36] The environmental issues resulting from oil and gas development are largely addressed under the Caspian Sea Convention in section 2 below, followed by the various water management issues regulated under various agreements in section 3. In the same way that there are parallels between the environmental issues in the Caspian Sea and southwestern Gulfs, considered in the previous chapter, so too are there similarities between the environmental issues in the Aral Sea Basin and the Mekong Basin, discussed in Chapter 4. Lessons learnt from the regulatory approach in both subregions may be of use in relation to regulating the water resource management issues in the Third Pole in particular, where dam-building activity is also growing at a considerable pace, and regulation may be found wanting. As will be seen in Chapter 5, the experience with the bilateral arrangements for the Indus River are currently, therefore, one of the few mechanisms in place in the Himalayas to manage water issues.

2. THE CASPIAN SEA CONVENTION AND ASSOCIATED PROTOCOLS

2.1 Geographical Context, Historical Background and Institutional Arrangements

The five state Parties to the Caspian Sea Convention are: Azerbaijan, Iran, Kazakhstan, Turkmenistan and Russia, with only Kazakhstan and Turkmenistan being central Asian states. The Caspian Sea itself, at 1000km-long, is the largest inland body of water on earth. Economic development, especially oil and gas exploration and exploitation, is a particular threat to the environment if not managed effectively. The United Nations Environment Programme (UNEP) confirms that the Caspian Sea, which is a partner of the Regional Seas Programme,[37] is a remnant of the ancient ocean Tethis, which previously connected the Atlantic and Pacific Oceans. One hundred and thirty rivers flow into the Caspian, the largest being the Volga. The long history and isolation of the Sea has resulted in impressive biodiversity, with more than 300

[36] Daene C McKinney, 'Cooperative Management of Transboundary Water Resources in Central Asia', in A Dinar, S Dinar, S McCaffrey and D McKinney, *Bridges over Waters: Understanding Transboundary Water Conflict, Negotiation and Cooperation* (World Scientific, Singapore, 2013) 187, 191.

[37] See <http://www.unep.org/regionalseas/programmes/independent/caspian/default.asp> accessed 14/11/16.

endemic species; the Caspian seal, for example, is one of only two freshwater seal species in the world, and of the important fisheries, the sturgeon, four species of which produce 90 per cent of the world's prized caviar, is most significant.[38] Extensive coastal wetlands host migrations for numerous birdlife. Yet the Sea has been heavily polluted, and a decade ago a Global International Waters Assessment (GIWA) for the Caspian Sea was completed, finding that:

> Habitat and community modification exerted the greatest impacts on the Caspian Sea region and was prioritised for Causal chain analysis and Policy option analysis … [These] concentrated on … loss of ecosystems or ecotones, and modification of … [them] including community structure and/or species composition.[39]

A similar approach has been taken by the UNECE.[40] The GIWA identified the most important root causes as: pollution from oil spills and agricultural discharges; introduction of invasive species; poaching of valuable species and an unsustainable fishery; and damming and regulation of rivers. The root causes were access to technology, especially old oil wells and pipelines; the availability of cheap out-of-date pesticides with no alternatives; the lack of facilities to decontaminate ships tanks

---

[38] Stephen J de Mora and Tim Turner, 'The Caspian Sea: A Microcosm for Environmental Science and International Cooperation' (2004) 48 *Marine Pollution Bulletin* 26, 28.

[39] F Stolberg, O Borysova, I Mitrofanov, V Barannik and P Eghtesadi, *Global International Waters Assessment – Caspian Sea*, GIWA Regional Assessment 23 (University of Kalmar/United Nations Environment Programme, 2006) 9.

[40] See 'Drainage Basin of the Caspian Sea', Chapter 4, in United Nations Economic Commission for Europe Convention on the Protection and Use of Transboundary Watercourses and International Lakes, *Second Assessment of Transboundary Rivers, Lakes and Groundwaters* (United Nations, 2011) 1–34; Libert and Lipponen comment on this: 'The transboundary waters were assessed for pressure factors, quantity and quality status, transboundary impacts, management responses, and future trends … Legal, institutional and socio-economic aspects are highlighted and cross-cutting themes that are challenges for the status of transboundary waters, such as climate change, are emphasized. By providing a picture of the state of the transboundary waters and by identifying joint priorities and challenges, the second assessment seeks to inform, guide, and stimulate further action by governments, river-basin organizations, the international community, and civil society'. See Bo Libert and Annukka Lipponen, 'Challenges and Opportunities for Transboundary Water Cooperation in Central Asia: Findings from UNECE's Regional Assessment and Project Work' (2012) 28(3) *International Journal of Water Resources Development* 565, 566.

and hulls; poor advice on fish quotas and inadequate laws; a failure to consider environmental factors when regulating dams; water extraction from rivers discharging into the Caspian Sea; and poor public participation and environmental awareness among stakeholders. Policy options fell into one of three groups: 'establishment and strengthening of regulations to control environmentally damaging activities in the region'; 'creation or refurbishment of facilities'; 'socio-economic actions'; and 'science and education'.[41] De Mora and Turner highlight the legacy of industrial contamination as a further major issue that has failed to be addressed in the subregion. Writing in 2004 they comment:

> The region has experienced major changes in industrial development in the past decade, with a collapse in the traditional heavy industry of the region. Environmentally, this has been a mixed blessing. Some areas have seen deterioration in environmental quality, as exemplified by the accumulation and deterioration of abandoned ships along the shoreline of Baku Harbour ... Alternatively, the downturn in economic activities has at least reduced the flux of contaminants to the local environment. Such an example is provided by the Sumgait industrial complex. First established 1946, a diverse range of industrial facilities was eventually developed at this location that included a pipe-production plant, a chlor-alkali works using mercury electrodes, a synthetic rubber plant, an aluminium smelter and several other chemical and petrochemical plants, together with two thermoelectric power plants. Poor environmental practices led to the region becoming one of the most polluted places in the world, with an estimated 5 million tons of waste accumulating in the vicinity of Sumgait.[42]

The historical background to the adoption of the Caspian Sea Convention has been a long and complex one. In more recent times the need to reach an agreement came to a head with the break-up of the Soviet Union. Thereafter, from the previous situation of just the Soviet Union and Iran having authority in the area, henceforth Russia – as the successor to the Soviet Union – and Iran were joined by newly independent Azerbaijan on the western shore, and Kazakhstan and Turkmenistan on the east. While agreements had been entered into between the Soviet Union and Iran in 1921 and 1940, the status of these agreements was unclear, despite the wish of Russia that they form the basis of any future arrangement. That arrangement was first and foremost to be based on the legal status of the Caspian Sea, which remains outstanding. Whether territory would be on the basis of the area being defined as a lake – and potentially subject to

---

41 Ibid, 9–10.

42 Above n. 38, 27.

international watercourse law[43] – or as an inland sea – and subject to any law of the sea regime[44] – has been argued vociferously.[45] The legal status question, while important background to the development of the *Caspian Sea Convention*, has however not prevented action in relation to the protection of the Caspian Sea itself, which is the focus here.

After the 1994 adoption of the *Almaty Declaration on Cooperation of the Environmental Protection of the Caspian Sea Region* in 2003, Azerbaijan, Iran, Kazakhstan, Russia and Turkmenistan (known as the 'Caspian Littoral States') signed the Convention.[46] The combined efforts of UNEP, the United Nations Development Programme (UNDP), the World Bank (WB) and others[47] contributed to this, reflecting its strategic

---

43 Convention on the Law of Non-Navigable Uses of International Watercourses (New York, 21 May 1997, by UNGA Res. 51/229), 36 ILM (1997) 700, in force 17 August 2014. There are eight Asian Parties to the New York Convention: Jordan, Lebanon, Palestine, Syria, Iraq, Qatar, Uzbekistan and Vietnam.

44 United Nations Convention on the Law of the Sea (Montego Bay, 10 December 1982) 1833 UNTS 3, 16 November 1994. There are over thirty Asian Parties to this: Armenia, Bahrain, Bangladesh, Brunei, China, Georgia, India, Indonesia, Iraq, Japan, Jordan, Kuwait, Laos, Lebanon, Malaysia, Maldives, Myanmar, Nepal, Oman, Pakistan, Papua New Guinea, Philippines, Qatar, South Korea, Russia, Saudi Arabia, Singapore, Sri Lanka, Palestine, Thailand, East Timor, Vietnam and Yemen.

45 Seyyed Rasoul Mousavi, 'The Future of the Caspian Sea after Tehran Summit' (2008–09) XXI(1–2) *The Iranian Journal of International Affairs* 27–42; Yusin Lee, 'Toward a New International Regime for the Caspian Sea' (2005) 52(3) *Problems of Post-Communism* 37–48; Kamyar Mehdiyoun, 'International Law and the Dispute over Ownership of Oil and Gas Resources in the Caspian Sea' (2000) 94(1) *The American Journal of International Law* 179–189; Yolbars A Kepbanov, 'The New Legal Status of the Caspian Sea is the Basis of Regional Cooperation and Stability' (1997–1998) II(4) *Perceptions – Journal of International Affairs* 5pp; and Sergei Vinogradov and Patricia Wouters, 'The Caspian Sea: Quest for a New Legal Regime' (1996) 9 *Leiden Journal of International Law* 94.

46 Shirin Akiner, 'Ten Years On: Achievements, Prospects, New Concerns', in Shirin Akiner (ed), *The Caspian: Politics, Energy and Security* (Routledge, Abingdon, UK and New York, USA, 2004) 337; in relation to security aspects, see also F Lebarre and M Tzankova, *Theoretical and Technical Perspectives on Security Sector Governance from the Adriatic to the Caspian Sea* (IOS Press, Amsterdam, 2011).

47 Although focused on the eastern part of the region, for detailed background and analysis of the recent situation, see UNEP, UNECE, UNDP, North Atlantic Treaty Organization (NATO), Regional Environment Centre (REC),

importance;[48] resource and other security issues have hence dominated scholarship in this area to date.[49] The Caspian Sea is geographically a cross-region and cross-subregion, as it is governed by the five littoral states which are located across different areas: Azerbaijan (Caucasus, Eurasia and western Asia), Iran (southwest Asia), Kazakhstan (central Asia), Turkmenistan (central Asia) and Russia (east Europe and north Asia). The strategic location of this area ensures that environmental issues in the Caspian Sea context relate both to energy security and environmental protection. As such, the objective of the *Caspian Sea Convention* is protection of the Caspian environment from pollution and sustainable use of biological resources.

The *Caspian Sea Convention* is supplemented by four Protocols: *Conservation of Biological Diversity* (*'Biodiversity Protocol'*),[50] *EIA in a Transboundary Context* (*'EIA Protocol'*),[51] *Regional Preparedness, Response and Cooperation in Combating Oil Pollution Incidents* (*'Regional Cooperation Protocol'*),[52] and the *Protection of the Caspian Sea against Pollution from Land-Based Sources and Activities* (*'Land-Based Sources and Activities Protocol'*).[53] While the Convention is in force, few of the Protocols have secured ratification by the littoral states; all of them need to do this before their entry into force. What can

OSCE, *Environment and Security in Europe, Transforming Risks into Co-operation: The Case of the Eastern Caspian Region* (2008).

48 See <http://www.tehranconvention.org/> accessed 12/3/15.

49 William Ascher and Natalia Mirovitskaya, *The Caspian Sea: A Quest for Environmental Security* (Kluwer, Dortrecht, 2000); Frederic Labarre and Maria Tzankova, *Theoretical and Technical Perspectives on Security Sector Governance from the Adriatic to the Caspian Sea* (IOS Press, Amsterdam, 2011).

50 Protocol for the Conservation of Biological Diversity (Ashgabat, 30 May 2014) unreported, not in force, only Turkmenistan has ratified (at 19/3/16). See generally, however, Convention on Biological Diversity (Rio de Janeiro, 22 May 1992) 31 ILM 822, in force 29 December 1993, which is the inspiration; numerous Asian states are Party to this.

51 Protocol on EIA in a Transboundary Context, 2012, unadopted, unreported, but available at: <http://www.tehranconvention.org/spip.php?article76> accessed 19/3/16. Similarly, see generally, Convention on Environmental Impact Assessment in a Transboundary Context (Espoo, 25 February 1991) 30 ILM (1991) 802, in force 27 June 1997 ('EIA Convention').

52 Protocol Concerning Regional Preparedness, Response and Cooperation in Combating Oil Pollution Incidents (Aktau, 12 August 2011) unreported, not in force, Kazakhstan still to ratify (at 19/3/16).

53 Protocol for the Protection of the Caspian Sea against Pollution from Land-Based Sources and Activities (Moscow, 12 December 2012) unreported, not in force, Kazakhstan and Russia still to ratify (at 19/3/16).

however be said is that interest in the resources of the Caspian region – particularly by China – means there is an urgent need for this to happen if the Caspian natural environment is not to be impacted further by ongoing development.

The institutional arrangements of the Convention are located in Part VI. Unlike the Regional Seas Conventions discussed in the previous chapter, there is no equivalent body to ROPME or PERGSA. Porkareh et al., who compare organizational arrangements under the Caspian Sea Convention with the Arabian Gulf Convention notes as recently as 2013 that both are 'under studied conventions'.[54] Instead, Article 22 establishes a Conference of the Parties (COP) consisting of representatives of the Parties. Its functions are enumerated in Article 22(10) and include: reviewing implementation and content (Article 22(10)(a)–(b)); adoption of protocols or amendments (Article 22(10)(c)); receiving and considering reports from the Parties and Secretariat (Article 22(10)(d)–(e)); and the establishment of subsidiary bodies (Article 22(10)(g)). The Secretariat is established under Article 23, with the typical administrative functions laid out in Article 23(4)(a)–(i).

2.2 Objectives, Principles and Procedures, including Implementation and Compliance

2.2.1 Convention

The Convention has the stated objective in Article 2 of 'the protection of the Caspian environment from all sources of pollution including the protection, preservation, restoration and sustainable and rational use of the biological resources of the Caspian Sea'. General obligations are found in Part II, Article 4: taking measures to prevent, reduce and control pollution (Article 4(a)); taking measures to protect, preserve and restore the environment (Article 4(b)); using resources so as not to cause harm (Article 4(c)); and cooperating with other littoral states and international organizations (Article 4(d)). Principles guiding action include the precautionary principle, the polluter pays principle, and the principle of access to information (Article 5). Parties also have a duty to cooperate in the development of protocols setting out measures, procedures and standards to implement the Convention, as noted above and outlined below (Article 6).

---

[54] Mohammad Hossein Porkareh, Seyed Masood Monavari, Yousef Molaei and Masood Akhavan Fard, 'Comparison of Regional Organisations for Protection of the Marine Environment (ROPME) and the Convention of Caspian Sea Provisions' (2013) 4(7) *Annals of Biological Research* 27, 27.

Part III contains measures for the prevention, reduction and control of pollution, with Article 7 addressing land-based sources. All appropriate measures are to be taken to prevent, reduce or control this (Article 7(1)); with additional protocols to include measures to prevent, reduce or control emissions at sources via 'low- and non-waste technology', and through licensing waste-water discharges 'based on promoting environmentally sound technology' (Article 7(2)(a)–(c)). Approaches to treatment for municipal waste water are to be 'step-by-step'; to reduce organic substances from industrial and waste water, 'the best available environmentally sound technology' is needed; the reduction of organic and hazardous substances from non-point sources must be dealt with by 'appropriate measures'; and coastal pollution must also be fully addressed (Article 7(2)(e)–(h)). Finally, where transboundary water pollution is likely to flow into the Caspian Sea, Parties must cooperate and establish joint bodies where appropriate (Article 7(3)).

Article 8 concerns pollution from seabed activities, Article 9 vessel pollution, Article 10 pollution from dumping, and Article 11 pollution from other human activities. Cooperation in the development of protocols is the key to the implementation of Articles 9–11, with vessel pollution to be regulated in accordance with the 'relevant international standards'. In relation to Article 10, the obligations to avoid dumping and to cooperate in the development of protocols are not applicable if a vessel or aircraft is threatened by complete destruction or loss, or is a danger to human or marine life, in which case dumping is accepted if it is the only way to avert the threat and it is likely that the damage will be less than it would be otherwise (Article 10(3)). The 'other human activities' contemplated by Article 11 include land reclamation and associated coastal dredging and dam construction (Article 11(1)); in all cases possible negative effects from sea-level fluctuations on the ecosystem must be mitigated by all appropriate measures (Article 11(2)).

Finally, Part III deals with invasive alien species in Article 12, and environmental emergencies in Article 13. In connection with invasive alien species – which are addressed in more detail in the *Biodiversity Protocol* below – Janusz makes reference to the definition of pollution in Article 1, 'the introduction by man, directly or indirectly, of substances or energy into the environment resulting or likely to result in such deleterious effects as harm to living resources and marine life, hazards to human health and hindrance to legitimate uses of the Caspian Sea'. She comments that this 'does not include biological alterations … the prevention of the introduction, control and combating of invasive alien species [is addressed in] a separate article because their introduction in the ecosystem is not generally regarded as pollution per se'.[55] The weakness of this separate article however is that 'it does not indicate any precautionary action regarding the alien species …'.[56]

In relation to emergencies, it is incumbent upon the Party of origin (of the emergency) to identify potentially hazardous activities that may result in such an emergency, and to notify others of this (Article 13(2)).[57] Procedures for EIA of such activities are briefly enumerated and the *Guidelines on Environmental Impact Assessment in a Transboundary Context* provide more detail, which are also directly relevant to the draft *EIA Protocol* which is considered below.[58] Part IV concerns protection, preservation and restoration of the marine environment in relation to the rational use of marine living resources (Article 14); coastal zone management (Article 15); and sea level fluctuation (Article 16). Part V contains procedures for EIA and cooperation. Article 17(1) requires measures for EIA for 'any planned activity … likely to cause significant adverse effect on the marine environment';[59] Article 17(2) contains additional requirements for the results of EIA to be disseminated; and Article 17(3) the development of protocols. Article 18(1) requires cooperation between Parties to be in accordance with international practice; Article 18(2) the formulation of an Action Plan; and Article 18(3) itemizes what is needed to achieve both. Article 19 contains provisions for monitoring; Article 20 research and development; and Article 21 exchange of, and access to, information.

Although the Convention incorporates a range of environmental controls and its adoption was a positive contribution to addressing the issues in the subregion, the framework nature of the *Caspian Sea Convention* is

---

[55] Barbara Janusz, 'The Framework Convention for the Protection of the Marine Environment of the Caspian Sea' (2005) 4(1) *Chinese Journal of International Law* 257, 263.

[56] Ibid, 264.

[57] See Peter Mark Taylor, Patricia Charlebois, Violeta Velikova and Frits Schlingemann, 'Oil Spill Preparedness and Co-operation in the Caspian Sea and Black Sea: Regional Successes and Lessons Learned', paper presented to the *International Oil Spill Conference* (2011).

[58] United Nations Environment Programme and Caspian Environment Programme, *Guidelines on Environmental Impact Assessment in a Transboundary Context in the Caspian Sea Region: Step by Step Procedures* (UNEP 2003).

[59] For detailed discussion of these provisions, their development, practice and law reform, see Simon Marsden, 'Developing Agreements for Transboundary Environmental Impact Assessment and Strategic Environmental Assessment in Asia', in Robin Warner and Simon Marsden (eds), *Transboundary Environmental Governance: Inland, Coastal and Marine Perspectives* (Ashgate Publishing, Farnham 2012) 141, 159–160.

noted by Janusz to be 'its greatest weakness'.[60] She summarizes as follows, noting the significance of implementation and compliance:

> Taken as a whole, the Framework Convention can be seen to mark a step forward in the littoral States' efforts to preserve the particularly fragile maritime environment of the Caspian Sea and to accomplish a general understanding of the issues affecting the Caspian Sea. The Framework Convention is a part of the progressive development of international legislation on environmental protection. However, the adequacy of the Framework Convention will be judged by its ability to protect the marine environment of the Caspian Sea.[61]

Provisions for implementation and compliance are found in Part VIII. Parties are required 'to designate a National Authority to co-ordinate implementation' (Article 26(1)), which is required to submit national implementation reports (NIRs) to the Secretariat (Article 27). Article 28 contains provision for the development of a Non-Compliance Procedure (NCP), for which to date there is no evidence. Article 29 requires Parties, 'taking into account relevant principles and norms of international law', to develop liability and compensation rules where the environment is damaged as a result of violations. In each case – Articles 26–29 – the provisions apply to both the Convention and its Protocols. Finally, Article 30 relates to dispute settlement in connection with the Convention, and requires this to be by 'consultations, negotiations or by any other peaceful means'. The Protocols under the Convention are considered in the next part.

2.2.2 Protocols

In relation to the first, the *Biodiversity Protocol*, the objectives are clearly established in Article 2: 'to protect, preserve and restore the health and integrity of the biological diversity and the ecosystem of the Caspian Sea as well as to ensure the sustainable use of biological resources'. These are furthermore elaborated in Article 2(a)–(c). The Convention applies to both the marine environment and associated 'wetlands of international significance' (Article 3), which are also protected by the *Convention on Wetlands of International Importance*.[62] The general provisions are

---

[60] Above n. 55, 258.

[61] Ibid.

[62] Also adopted in Iran; see the Convention on Wetlands of International Importance, especially as Waterfowl Habitat (Ramsar, 2 February 1971) 11 ILM 963, in force 21 December 1975. This has again been ratified by numerous Asian Parties.

located in Article 5, and oblige taking 'all appropriate measures to protect, preserve and restore the marine environment' (Article 5(a)); using natural resources so as 'not to cause harm to the marine environment' (Article 5(b)); emphasizing threatened species (Article 5(c)) and 'areas that are unique, highly sensitive or regionally representative ... notably by the establishment of protected areas' (Article 5(d)); adoption of strategies, action plans and programmes, and implementation of legislation (Article 5(e)); monitoring (Article 5(f)); identifying and compiling inventories (Article 5(g)); ensuring consistency in sectoral strategies and action plans (Article 5(h)); applying integrated management (Article 5(i)); controlling pollution sources (Article 5(j)); and cooperation among one another, and with international organizations (Article 5(k)).

Part II of the *Biodiversity Protocol* concerns the protection and conservation of species, and Part III, protected areas. In relation to the first, Article 6 outlines the measures required for species protection; Article 7 regulates alien species; and Article 8 genetically modified organisms. Of the second, Article 9 contains provisions for the designation of protected areas, including those of a transboundary nature (Article 9(2)); Article 10 for their management; and Article 11, the procedures for their establishment and listing. Annex I and II supplement these obligations with categories for identifying threatened species, and through compiling an IUCN-inspired 'Caspian Red Book' (I) and common criteria for inclusion on the List of Protected Areas of the Caspian Sea (PACS List) (II) outlined. In addition to the inspiration of the *Convention on Biological Diversity* (CBD),[63] there are other biodiversity conventions indirectly referred to in these provisions, such as Article 6(c) and (f); see for example the *Convention on International Trade in Endangered Species* (CITES)[64] and the *Convention on the Conservation of Migratory Species* (CMS).[65] This is emphasized in various provisions, where the words 'in conformity with applicable international treaties' are used, such as Article 10(1) and (2).

Part IV sets out instruments and requirements for the protection and conservation of both species and protected areas. Article 12 is focused upon the role of Integrated Coastal Zone Management (ICZM); Article 13 EIA; Article 14 access to genetic resources; Article 15 access to and

---

63 Above n. 50.

64 Convention on International Trade in Endangered Species of Wild Fauna and Flora (Washington DC, 3 March 1973) 993 UNTS 243, in force 1 July 1975.

65 Convention on the Conservation of Migratory Species of Wild Animals (Bonn, 23 June 1979) 19 ILM 15, in force 1 November 1983.

transfer of technology; Article 16 scientific and technical cooperation and assistance; Article 17 exchange of information; Article 18 environmental education and public awareness; and Article 19 contains the reporting mechanism. In relation to the latter, an NIR is required on the status of biodiversity, habitats, protected areas included on the PACS List, and the implementation of environmental education and public awareness programmes. Dispute settlement is contained in the final clauses of the Protocol, Article 22, which simply states that disputes are to be 'settled in accordance with Article 30 of the Convention'.

The requirement for coastal zone management found in the *Biodiversity Protocol* (Article 12) is an important one, and builds on the related obligation in the *Caspian Sea Convention* (Article 15). The Convention provision cited above requires states to 'take necessary measures to develop and implement national strategies and plans for planning and management of the land affected by proximity to the sea'. The Protocol provision gives examples of the application of these 'national strategies and plans'. Reviewing practice in the early years of implementation, Nouri et al. describe the coastline policies based on a 'strengths and weaknesses' approach as follows, with eight examples of the approach taken by the littoral States:

> 1. equalization of distribution of population and activities along the coast consistent with the existing ecological capacities, 2. strengthening potential tourism capabilities and promoting its management efficiency at the national and international levels and giving priority to the private sector, 3. preventing irregular expansion of coastal cities and considering the shoreline's limitations for establishing new human settlements, 4. preventing marine environment pollution, prevention of forest destruction, protection of soil-water resources and enhancing shoreline ecotourism via strengthening environmental management ... 5. expansion of fishing ports, fishing fleet, fishing industries and increasing the aquaculture production and export, 6. development and equipment of marine terminals and ports and providing required facilities to meet with the increasing demand for cargo transit, 7. land allocation consistent with coastal environment's potential's and capabilities and 8. appropriate utilization of soil water resources to increase quality of agricultural products.[66]

Pak and Manuchehr furthermore consider Iran's experience with ICZM in the Caspian Sea, Arabian Gulf and Gulf of Oman. Distinguishing the Caspian Sea from the two Gulfs, they consider the needs of the following

---

[66] J Nouri, AR Karbassi and S Mirkia, 'Environmental Management of Coastal Regions in the Caspian Sea' (2008) 5(1) *International Journal of Environmental Science and Technology* 43, 46–47.

plans in the former: environmental management plan, spatial plan for the coastal zone, public participation plan, organizational/institutional structure plan, and a monitoring and evaluation plan. They comment:

> Studies are currently underway and the above-mentioned plans are to become complete in 2007. Governmental concerns regarding coastal problems have been on the agenda of several meetings among the central/regional authorities as well as the coastal scientists and specialists which have led to some basic findings that have already shed their light to the continuation of Iran's ICZM studies and efforts. Till now the basic findings are as follows:
>
> - In order to solve the coastal problems, the existing management structure for coastal areas requires fundamental change or major re-engineering.
> - Coastal boundaries and limits must be carefully defined and marked on the ground in order to have control on the enforcement of the proposed regulations.
> - For materializing the sustainable management a number of strategic and executive plans are required.
> - Transparency and public awareness are necessary for making ground to have the private sector and NGOs involved in the process of ICM implementation.
> - New laws and regulations pertaining to coastal issues should be enacted.
> - Cooperation with neighboring countries at the north and at the south is of prime importance.[67]

In relation to the second subsidiary agreement, the draft *EIA Protocol* draws strongly on the *EIA Convention*,[68] in that it also operates in a transboundary context.[69] In connection with this, the link with access to information and public participation in environmental matters is also acknowledged in the recitals (Recital 5).[70] At the Fifth Meeting of the Parties of the *Caspian Sea Convention* in 2014, the importance of these

---

[67] Above n. 3, 769.

[68] Above n. 51.

[69] See also Marsden and Brandon, Chapters 4 and 8, above n. 19; and Marsden above n. 59.

[70] This acknowledges the role of the Convention on Access to Information, Public Participation in Decision-Making and Access to Justice in Environmental Matters (Aarhus, 25 June 1998) 2161 UNTS 447, in force on 30 October 2001. See Marsden and Brandon, Chapters 3 and 8, above n.19; and for further discussion, see Simon Marsden, 'Public Participation in Transboundary Environmental Impact Assessment – Closing the Gap between International and Public Law?', in Brad Jessup and Kim Rubenstein (eds), *Environmental Discourses in Public and International Law* (Cambridge University Press, Cambridge, 2012) 238.

matters was emphasized throughout the provisions of both the Convention and each of the Protocols.[71] The Protocol also implements Articles 7, 17 and 18 of the *Caspian Sea Convention* as indicated (Recital 2). Furthermore, as seen in the evidence of practice section below, it also draws on the former non-binding *Guidelines*[72] that preceded the *EIA Protocol*. The stated objective of the Protocol is found in Article 2: 'to implement effective and transparent EIA procedures in a transboundary context to any proposed activity which is likely to cause significant transboundary impact on the marine environment …'. The terms are defined in Article 1, including 'Party of Origin', the state responsible for the potential transboundary impact, and 'Affected Party', the receiving state.

The general provisions are found in Article 4. These require the adoption of measures to implement the Protocol 'with respect to proposed activities listed in Annex I, that are likely to cause significant transboundary impact as determined in accordance with the criteria set forth in Annex II … [and] that permits public participation and preparation of the … [EIA] documentation described in Annex III …' (Article 4(1)). In accordance with the usual procedure, the EIA procedure is to be applied 'prior to a decision to authorize or undertake a proposed activity' (Article 4(2)); and an 'equivalent opportunity to the public of the Party of Origin and the public of the Affected Party to participate' is required (Article 4(5)).[73] Article 5 requires the Party of Origin to notify where proposed activities may have potentially significant environmental effects. A wide range of information is to be included in this (Article 5(2)); it must be provided in a language that is understood (Article 5(3)); a response is required from the Affected Party as to whether it wishes to participate in the procedure (Article 5(5)); and if so, it must provide

---

[71] Framework Convention for the Protection of the Marine Environment of the Caspian Sea, Conference of the Parties' Fifth Meeting [29–30 May 2014, Ashgabat] Turkmenistan, Items 5 and 12 of the Provisional Agenda, *Recommendations and Analysis Based on the Comments by the Aarhus Convention Secretariat on Public Participation under the Tehran Convention*, Note by the Interim Secretariat, TC/COP5/INF.6

[72] Above n. 58.

[73] This is the non-discrimination principle, which may however preference certain groups; see Simon Marsden, 'Enforcing Non-Discrimination in Transboundary Environmental Impact Assessment: Advantages for EU Citizens from the Transposition of the Espoo and Aarhus Conventions?' (2009) 6 *Journal for European Environmental and Planning Law* 437.

certain information (Article 5(7) and (8)). Article 6 requires communication between the competent authorities of the Parties as to language, detailed arrangements and time-frames.

In connection with the EIA documentation, Article 7 governs the preparation and transmittal of draft materials. The proponent is responsible for preparation of the draft documentation 'for the purposes of public consultations and review by the competent authority of the Affected Party (Article 7(2)). Article 8(1) obliges the Origin and Affected Parties to ensure the public are informed about the proposal, documentation and consultation opportunities. The documentation is to be made available to the public (Article 8(2)), who must be given the opportunity to comment (Article 8(3)); with the competent authority of the Affected Party reviewing and commenting on the draft documentation (Article 8(4)); finally, the Party of Origin must ensure the proponent incorporates comments received into the final documents, with explanation of how they were taken into account (Article 8(5)(c)).

Before the final decision is made, consultations between the Origin and Affected Parties must take place before the final decision, if the Affected Party wishes to discuss measures to reduce the potential effects (Article 9(1)). In accordance with Article 10(1), it is the responsibility of the Party of Origin to take the comments received further to Article 8 into account when reviewing the final documentation and making the final decision; Article 10(2) furthermore requires reasons and considerations on which the decision has been taken to be provided; and Article 10(3) that information on the final decision be made available to those who contributed to the process. Other provisions in the Protocol concern post-project analysis (Article 11); implementation reporting (Article 12); review and amendment of the Protocol, and duties of the Secretariat (Article 13); together with dispute settlement, which is to be in accordance with the provisions of the Convention (Article 15).

The third protocol is the *Regional Cooperation Protocol.* Both this, and the fourth, the *Land-Based Sources and Activities Protocol*, are similar to those found in connection with subsidiary agreements with the same or similar names under the *Arabian Gulf/Gulf of Oman Convention* and *Red Sea/Gulf of Aden Conventions* as considered in the previous chapter. In connection with the *Regional Cooperation Protocol*, these other arrangements are acknowledged in the recitals (Recitals 7 and 9), and the objective (Article 3) is therefore to establish regional measures in relation to protection against oil pollution from various activities. Unlike the *Arabian Gulf/Gulf of Oman Convention*, environmental emergency is not defined in the Protocol, but it is included in the activities which are its focus, with the recitals indicating 'that special measures are necessary in the event of oil pollution incidents originating from ships, pipelines, fixed and floating platforms, and abandoned wellheads and land-based sources of pollution' (Recital 2). Oil pollution incidents and accidents are however defined as: '... an occurrence or series of occurrences having the same origin, which results or may result in a discharge of oil and which poses or may pose a threat to the marine environment, or to the coastline, and which requires emergency action or other immediate response' (Article 1(a)).

General provisions are contained in Article 4, which require the Parties to take measures 'to prepare for and respond to' incidents, establish guidelines for action, establish a regional mechanism (like ROPME or PERGSA, or specifically MEMAC as considered in Chapter 2) and initiate a Caspian Sea Cooperation Plan. In relation to the regional mechanism, this is designed to assist Parties 'in reacting promptly and effectively to oil pollution incidents' (Article 13(1)); its functions include establishing close relations with national authorities and relevant international and regional bodies, coordinating regional activities such as technical cooperation and training, collecting and disseminating information, preparing procedures for information exchange, acting as a focal point, and making proposals for updating the Caspian Sea Plan (Article 13(2)).

Parties are furthermore required to: establish national systems and contingency plans for combating oil pollution incidents, identifying authorities and contact points (Article 5(1)); prepare and implement a national contingency plan for 'preparedness and response' (Article 5(2)); establish and maintain equipment to effectively deal with oil discharges (Article 5(3)); and set up training programmes to develop capacity (Article 5(4)). Article 6 concerns the dissemination and exchange of information, with NIR obligations to the Secretariat and other Parties. Practical measures concerning pollution-reporting procedures for those on ships, offshore units, sea ports and oil handling facilities (Article 7), operational measures in the event of an oil spill (Article 8), and emergency plans on board ships, offshore units, in ports and at oil-handling facilities (Article 9), follow. The latter provision makes direct reference to relevant international law in this area.[74] Additional measures govern assistance requests from the other Parties (Article 10), and

[74] Namely the International Convention on Oil Pollution Preparedness, Response and Cooperation (London, 30 November 1990) 1891 UNTS 51, in force 13 May 1995; and the International Convention for the Prevention of Pollution from Ships (London, 2 February 1973) 12 ILM 1319, in force 12 October 1983.

reimbursement of the costs of assistance (Article 11). Article 12 contains institutional provisions, which requires the COP to, among other things: keep under review the implementation and content of the Protocol; consider and adopt amendments; consider NIRs; and establish subsidiary bodies, which would include any NCP. Disputes are to be settled in accordance with Article 30 of the Convention (Article 16).

Finally, the *Land-Based Sources and Activities Protocol* has the stated purpose: 'to prevent, control, reduce and to the maximum extent possible eliminate pollution' (Article 1), 'conscious of the serious danger posed to the marine environment and coastal areas, living resources and human health by pollution from land-based sources and activities' (Recital 6). After defining numerous terms relevant for pollution control in Article 2, such as 'best available techniques', 'best environmental practice', 'hot-spot', 'point sources', 'diffuse sources', 'emission controls', 'emission limit values', 'environmental quality standard' and 'environmental quality objective', the Protocol defines in detail the scope of application in Article 3, which considers both the origin and the destination of the pollution in question.

The general obligations are found in Article 4(1), which require Parties to 'take all appropriate measures … to prevent, control, reduce and to the maximum extent possible eliminate pollution'. In doing so, Parties are to: apply the precautionary and polluter pays principles; promote cooperation between and among themselves; ensure environmental factors (including health) are taken into account in the development of plans and programmes; take preventative measures where needed; and take special measures where the spawning grounds of valuable species (in particular sturgeon) may be affected (Article 4(2)). National authorities must be designated to coordinate implementation (Article 5(1)); and in implementing the Protocol, Parties must adopt plans and programmes based on pollution-source control with timetables for completion, addressing the activities and substances that are listed in Annex I via emissions controls, timetables and using or promoting best available techniques and taking into account factors defined in Annex V (Article 5(2)).

Common guidelines and standards must be formulated and adopted (Article 6); pollution from point and diffuse sources – the latter focused heavily on agriculture – regulated (Articles 7 and 8); and pollution from other activities – with concern expressed for effects as a result of sea-level fluctuations – prevented (Article 9). National action plans are a necessary component (Article 7(2)); these must identify pollutant hot-spots, and be subject to review and revision by the COP every two years. Authorization or regulation is required for point source emissions, taking account of Annex IV and other matters (Article 7(3)); and regular inspection to assess compliance is to be established by the Parties (Article 7(4)). An integrated approach to coastal area development is needed 'to reduce negative impacts of natural hazards such as long-term sea-level fluctuation, storm-surges, storms, earthquakes and coastal erosion on the population and infrastructure of the coastal areas' (Article 10(1)); furthermore, 'Parties shall take all appropriate measures to reduce and reverse deforestation and land degradation' (Article 10(2)).

Transboundary impacts and EIA generally are covered by Articles 11 and 12 respectively. In connection with the first, information and consultation based thereon is required where there are likely significant effects from land-based sources of pollution originating in one Party and potentially impacting another (Article 11(1)); where the discharge is from a watercourse the establishment of joint bodies must form the basis of any cooperation (Article 11(2));[75] and liability will not flow from pollution originating in the territory of non-Parties that flows between the territory of Parties (Article 11(3)). In connection with EIA generally, Article 12(1) makes reference to Article 17 of the *Caspian Sea Convention* in requiring the development and adoption of EIA guidelines;[76] Article 12(2) requires the introduction and application of EIA procedures; and Article 12(3) that planned activities be subject to prior written authorization, taking 'fully into account the findings and recommendations'. Experience with transboundary EIA in this region was reported on in 2007,[77] which concerned offshore oil development by Azerbaijan and Turkmenistan. Since then, China has been involved in a number of oil and gas projects, which are commented on in the penultimate section of this chapter.

Article 13 concerns information and data collection, monitoring and assessment; Article 14 deals with exchange of information; and Article

---

[75] The establishment of joint bodies is a requirement of international water law in some cases. See for example Article 9(2) of the Convention on the Protection and Use of Transboundary Watercourses and Lakes (Helsinki, 17 March 1992) 31 ILM (1992) 1312, in force 6 October 1996. For discussion see Chapter 6 in Marsden and Brandon, above n. 19; also Simon Marsden, 'The Helsinki Water Convention: Implementation and Compliance in Asia' (2015) 20(2) *Nordic Environmental Law Journal* 119–129.

[76] Above n. 58.

[77] See Rie Tstsumi and Kristy Robinson, 'Environmental Impact Assessment and the Framework Convention for the Protection of the Marine Environment of the Caspian Sea' in Kees Bastmeijer and Timo Koivurova, *Theory and Practice of Transboundary Environmental Impact Assessment* (Martinus Nijhoff, Leiden, 2007).

15 public participation. These provisions are however limited in scope and apply primarily between the Parties. Article 13(1) for example limits information collection to the monitoring programmes, and the exchange of information is aimed at inter-Party exchange under Article 14. Similarly, public participation – which is heavily dependent upon access to information that may not be made available under Article 15(2) – is only 'promoted' under Article 15(1), and curiously makes reference to Article 21 on dispute settlement, which is not an avenue for the public to pursue in relation to grievances as it is limited to state Parties. Nonetheless, Article 15(2) does require Parties to 'facilitate public access to the information concerning conditions of the marine environment … measures taken or planned to be taken to prevent, control and reduce pollution taking into account provisions of existing international agreements concerning public access to environmental information which they are Party to'. This indicates that while the Convention and Protocol may not enable access to information in relation to EIA for example, membership of other arrangements may do so, notably the *Convention on Access to Information, Public Participation in Decision-Making and Access to Justice in Environmental Matters.*[78]

Article 16 concerns cooperation and assistance, Article 17 reporting, and Article 18 compliance. The NIR requirements in Article 17 are detailed, and a very positive aspect of this Protocol, requiring reports to be submitted to the Secretariat on implementation measures which must include a varied range of matters. In Article 17(2) these include: (a) 'information on legal and regulatory measures, action plans, programmes and other steps'; (b) Annex I substances of concern discharged; (c) statistical data on point source authorizations; (d) information from the monitoring process; (e) information 'as a result of activities that alter or may alter the coastline or that adversely affect habitats'; (f) hotspot information on elimination or control; and (g) information on implementation of action plans, programmes and measures. Article 18(1) enables the development of an NCP by the Parties; with the COP reviewing and evaluating compliance on the basis of the NIR requirement (Article 18(2)(a)). Measures to ensure compliance are available under Article 18(2)(b), although these are not specified. Finally, the COP is empowered to review implementation and compliance further to the institutional provisions in Article 19(1), which also include reviewing the content of

---

[78] Commonly known as the 'Aarhus Convention', see above n. 70. Of the Caspian Littoral States, however, neither Iran nor Russia are either a Party or signatory to this treaty. See Marsden and Brandon, above n. 19, Chapter 3.

the Protocol; programmes, plans and measures; guidelines standards and criteria; information exchange; the list of hot-spots; and the establishment of subsidiary bodies, which would include a formal NCP. This has yet to be established, as with the Convention.

3. WATER RESOURCE MANAGEMENT IN THE ARAL SEA BASIN: THE JOINT MANAGEMENT AND CONSERVATION, INTERNATIONAL FUND AND INTERSTATE COMMISSION AGREEMENTS

3.1 Geographical Context, Historical Background and Institutional Arrangements

The Aral Sea lies to the east of the Caspian Sea between Kazakhstan and Uzbekistan. In the last fifty years it has lost half its surface area and three-quarters of its volume because of the irrigation policies of the former Soviet Union and the ineffectiveness of the response of the central Asian states since independence, which have not diversified their economies sufficiently to improve the situation. The Basin ranges topographically from the extensive Turanian plains in the west, to the mountain ranges of the Pamirs and Tien Shan in the east, ranges which, as noted above, are part of the Third Pole as discussed in Chapter 5 of this book. The Sea is now effectively divided in two because of the lack of water from the Amu and Syr Darya (rivers), which run from the mountains through the desert to the Aral Sea.[79]

Extinction of native fish species, and consequences for the associated commercial fishery, is but one of the negative environmental effects that have resulted, and Campins Eritja et al. argue that the regulation of the fishery in the Basin should potentially be considered in connection with international fisheries law.[80] The salinity of the Basin has furthermore impacted drinking water supplies, as well as killing the other ecosystems of the region. The degradation of vegetation – and deforestation – has

---

[79] See for example, McKinney, above n. 36, 189–194. See also generally for background, M Glantz, *Creeping Environmental Problems and Sustainable Development in the Aral Sea Basin* (Cambridge University Press, Cambridge, 1999).

[80] Mar Campins Eritja, Jaume Saura Estapá and Xavier Pons Ráfols, 'Towards Improved Regional Co-operation over Water Uses in Central Asia: The Case of Hydroelectric Energy and Inland Fisheries' (2016) 6(1) *Asian Journal of International Law* 119, 144–157.

exposed land to erosion, which has been exacerbated also by overgrazing. In addition, a legacy of industrial and radioactive pollution continues to this day, potentially causing significant added problems. The water resource management problem and associated environmental impacts in the subregion arise because of the divergent needs of the upstream and downstream states, as McKinney clearly summarizes:

> The Central Asian republics depend on the rivers of the Aral Sea Basin for drinking water, irrigation and hydroelectric power. In the upstream countries of the Basin, Kyrgyzstan and Tajikistan, the rivers are used for hydroelectric power, especially during winter months, while downstream, in Turkmenistan, Kazakhstan and Uzbekistan, they are used for agricultural purposes in the summertime. The post-independence upstream shift in water use away from irrigation has created disputes between the upstream and downstream countries over how the region's transboundary waters should be managed. Successful cooperative sharing of water and other natural resources is essential for the long-term prosperity and security of the region.[81]

Following the dissolution of the Soviet Union, the central Asian states knew that stabilizing transboundary water relations was the key to future cooperation, as they were heavily dependent upon irrigated agriculture. Lim explains that the inheritance of Soviet water allocation regimes is the main reason for transboundary water disputes in present-day central Asia. These water management rules prioritize irrigated agriculture but do not take into account the current power generation needs of the upstream states of Kyrgyzstan and Tajikistan.[82] Although the downstream riparians continue to benefit from generous water allocations, they now became dependent on their upstream neighbours for these, increasing the possibility of conflict.

To avoid this, on 12 September 1991 the water ministers of each republic declared that joint water resources management would be established on the basis of equity and mutual benefit. Eventually, the *Interstate Water Resources Agreement* between Kazakhstan, Kyrgystan, Tajikistan, Turkmenistan and Uzbekistan was concluded to replace the

---

[81] McKinney, above n. 36, 187.

[82] Michelle Lim, 'Effective Governance across Central Asian Boundaries: What Lessons does the Central Asian Transboundary Watercourse Management Experience Provide for Mountain Biodiversity Conservation across the Kyrgyz-Tajik Boundary?', in Robin Warner and Simon Marsden (eds), *Transboundary Environmental Governance: Inland, Coastal and Marine Perspectives* (Ashgate Publishing, Farnham, 2012) 111, 119.

former centralized system, which had resulted in significant conflict.[83] Libert and Lipponen refer to this as the 'key regional agreement';[84] this appears to be because it was the first of the agreements between the central Asian states, and also because it was one of the few of the agreements to include all five central Asian states.[85] This was a framework for annual negotiations which would be regulated by protocols. The downstream states of Kazakhstan and Uzbekistan were therefore to pay for electricity produced by the Toktogul Dam upstream, which was of particular benefit to Kyrgyzstan. However, as no agreement was reached with Tajikistan, it failed to benefit from the arrangement, which as a result was mired in disputes about the cost of the water to be supplied. This 1992 Agreement failed to provide for the winter energy supplies of Kyrgyzstan and Tajikistan and resulted in their turning to hydroelectricity to replace the previous Russian-provided energy.[86]

Further to the *Interstate Water Resources Agreement* and following a conference which considered the problems of the Aral Sea region, in 1995 the *Nukus Declaration* was signed.[87] This emphasized the willingness of the five states to cooperate and establish bodies to address the problem.[88] However, notable exclusions were Afghanistan and Iran, the

---

[83] See Jason Bart, 'Weaponizing Water: Water and Energy as Sources of Conflict among the Central Asian Soviet Successor States' (2013) 22(1) *Michigan State International Law Review* 409; and Christine Bichsel, 'Liquid Challenges: Contested Water in Central Asia' (2011) 12(1) *Sustainable Development Law and Policy* 24.

[84] Above n. 40, 568.

[85] The others with comprehensive membership are the *Saving the Aral Sea Agreement, Interstate Commission Statute* and the *Agreement on Joint Actions for Addressing the Problems of the Aral Sea and its Coastal Area, Improving the Environment, and Ensuring the Social and Economic Development of the Aral Sea Region.* For details of the latter, see below n. 90.

[86] Bo Libert, Erkin Orolbaev and Yuri Steklov, 'Water and Energy Crisis in Central Asia' (2008) 6(3) *China and Eurasia Forum Quarterly* 9, 11.

[87] Resolution of the Heads of States of Central Asia on Work of the EC of ICAS on Implementation of Action Plan on Improvement of Ecological Situation in the Aral Sea Basin for the 3–5 Years to Come with Consideration for Social and Economic Development, Concerning Water Quality in the Areas of the Aral Sea, the Amu Darya Basin, and the Syr Darya Basin, Nukus Declaration (Nukus, 20 September 1995). See <http://www.internationalwaterlaw.org/documents/regionaldocs/nukus_declaration_eng.pdf> accessed 14/11/16.

[88] Joseph MacKay, 'Running Dry: International Law and the Management of the Aral Sea Depletion' (2009) 28(1) *Central Asian Survey* 17, 22–23.

territories of which include the upper watershed of the Amu Darya.[89] Various agreements followed between all or some of the central Asian states concerning water and other relevant matters.[90] In April 1996, for example, an agreement was reached between Kazakhstan, Kyrgyzstan and Turkmenistan over resources trade;[91] it followed an agreement in January 1996 between Turkmenistan and Uzbekistan, whereby the latter paid the former rent for pumping stations and a water storage area.[92] In addition and of some importance, were the *Saving the Aral Sea Agreement*, which was concluded in 1999;[93] and the above-mentioned *Interstate Commission Agreement*, almost a decade later in 2008.[94]

The year 2007–2008 was a pivotal one in central Asia and drew attention to the divergent needs of the upstream and downstream states in a very clear way; it was because of the events of this time that the 2008

---

[89] See Erika Weinthal, *State Making and Environmental Cooperation: Linking Domestic and International Politics in Central Asia* (MIT Press, Cambridge, 2002), who questions what would happen if Afghanistan were to claim the headwaters in the interests of economic development.

[90] Agreement on Joint Actions for Addressing the Problems of the Aral Sea and its Coastal Area, Improving the Environment, and Ensuring the Social and Economic Development of the Aral Sea Region (Kzyl-Orda, 26 March 1993) – between Kazakhstan, Kyrgystan, Tajikistan, Turkmenistan and Uzbekistan; Agreement on Cooperation over Water Management Issues (Chardjev, 16 January 1996) – between Turkmenistan and Uzbekistan; Agreement on the Use of Fuel and Water Resources, Construction and Operation of Gas Pipelines in Central Asian Region (Tashkent, 5 April 1996) – between Kazakhstan, Kyrgystan and Uzbekistan; Agreement on the Use of Water and Energy Resources of the Syrdarya Basin (Bishkek, 17 March 1998) – between Kazakhstan, Kyrgystan, Tajikistan and Uzbekistan; Agreement on Cooperation in the Area of Environment and Rational Nature Use (Bishkek, 17 March 1998) – between Kazakhstan, Kyrgystan and Uzbekistan; Agreement on the Main Principles of Interactions in the Field of Rational Use and Protection of the Transboundary Watercourses of the CIS (Moscow, 11 September 1998) – between Belarus, Russia, Kazakhstan and Tajikistan; and Agreement on the Parallel Operation of the Energy Systems of Central Asian States (Bishkek, 17 June 1999) – between Kazakhstan, Kyrgystan, Tajikistan and Uzbekistan.

[91] Agreement on the Use of Fuel and Water Resources etc., ibid; see MacKay, above n. 88, 23.

[92] Agreement on Cooperation over Water Management Issues, above n. 90; Kai Wegerich, 'Water Resources in Central Asia: Regional Stability or Patchy Make-Up?' (2011) 30(2) *Central Asian Survey* 275, 282.

[93] Above n. 7. See: <http://ec-ifas.waterunites-ca.org/news/54-the-intergovernmental-agreement-on-the-status-of-the-international-fund-for-saving-the-aral-sea-is-ratified.html> accessed 14/11/16.

[94] Above n. 8.

arrangements were finally put in place. A very cold winter was followed by a dry spring and summer. Heavy use of hydro-electricity by Kyrgyzstan during the winter led to low water flows to the Syr Darya, which had dire implications for Uzbekistan and Kazakhstan in relation to their seasonal irrigation needs. Tajikistan was also impacted heavily, with heating and water supplies unavailable for long periods at a time. Libert et al. explain the water resource issues in detail, and highlight the dam-building activities that were seen by some states as the cause of the crisis, and by other states as the solution to this state of affairs.[95]

3.2 Objectives, Principles and Procedures, including Implementation and Compliance

As indicated above, the central Asian states agreed to maintain the basic principles of water allocation that prevailed in Soviet times, with the *Interstate Commission Agreement* acknowledging the equal rights of member states to use the interstate resources and their responsibility to protect them. It therefore affirmed the continuation of existing Soviet structures and principles of interstate water allocation. The issues of concern are addressed to a large extent by the UNECE *Water Convention*,[96] and the UN *Watercourses Convention*;[97] however although both are particularly relevant in the subregion,[98] effectiveness depends upon ratification as well as implementation and compliance, which is yet to become a comprehensive reality.

The *Saving the Aral Sea Agreement* established the International Fund for Saving the Aral Sea (IFAS) under Article 1, which, among other bodies, includes the Interstate Commission for Water Coordination (ICWC) from the *Interstate Commission Agreement.* Article 2 provides the IFAS legal personality, and Articles 3–12 provide privileges and immunities of the various organizations and personnel. The *Interstate*

---

[95] Libert et al., above n. 86.

[96] Above n. 75. Kazakhstan, Turkmenistan and Uzbekistan are Parties to this treaty.

[97] Above n. 43. See Flavia Rocha Loure and Alistair Rieu-Clarke (eds), *The UN Watercourses Convention in Force: Strengthening International Law for Transboundary Water Management* (Routledge, Abingdon, 2013). Uzbekistan is the only central Asian Party to this treaty.

[98] Marsden, above n. 75. The role of the Desertification Convention is also significant: see UN Convention to Combat Desertification in Those Countries Experiencing Serious Drought and/or Desertification, Particularly in Africa (Paris, 14 October 1994) 1954 UNTS 3, in force 26 December 1996. For discussion of relevance see MacKay, above n. 88, 21–22.

Commission Agreement confirms the creation of the ICWC according to the *Saving the Aral Sea Agreement*, and outlines the structure for this in Part III. This part states that ICWC meetings are held quarterly on a rotating basis by the member states (section 3.2); decisions are made by consensus (section 3.4); and that executive bodies are to be established to fulfil particular tasks (section 3.7).

Article 7 of the *Interstate Commission Agreement* established the ICWC, which was intended to control, ensure the rational use of, and protect interstate water resources. It was also responsible for the management, allocation and development of regional management polices to ensure the rational utilization of water resources; furthermore it governed the activities of the two regional water organizations for the Amu and Syr Darya basins. From 1992, the ICWC was responsible not only for development and allocation planning, but also quality control, conservation and environmental protection. Although Libert et al. refer to the ICWC as 'an important structure for the resolution of operational water distribution issues on the Amu Darya and Syr Darya under the 1992 Agreement',[99] they are very clear about its limitations:

> Only water authorities are represented in ICWC, making it difficult to apply an intersectoral approach for cooperation involving, in addition to irrigation, energy and protection of ecosystems. This is also a hindrance to implementing its principal task – enforcing the implementation of water allocation in compliance with the 1992 Agreement and preparing new rules and procedures for sharing the waters of the transboundary rivers of Central Asia.[100]

The executive bodies may include the Secretariat, Basin water organizations for the two main watercourses, a Scientific Information Centre, a Coordination Metrological Centre, and a Training Centre (section 5.1). Working groups may also be established periodically (section 5.2), and the roles of the Secretariat (section 5.3), Basin Water Organizations (sections 5.4 and 5.5), the Scientific Information Centre (section 5.6), the Coordination Metrological Centre (section 5.7) and the Training Centre (section 5.8) are all outlined. Ziganshina also discusses the role of the ICWC, highlighting that a 1993 *Agreement on Joint Actions for Addressing the Problems of the Aral Sea and its Coastal Area, Improving the Environment, and Ensuring the Social and Economic Development of the Aral Sea Region*[101] placed the ICWC underneath the newly established

---

99 Libert et al., above n. 86, 18.

100 Ibid.

101 Above n. 90.

Interstate Council on the Aral Sea (ICAS), subsequently to be transformed into the IFAS. The significance of this for Ziganshina is that:

> Although a compliance control system is yet to be established in the region, the activities of the ICWC and its executive bodies appear to provide a basis on which such a system can be built. The ICWC does not have a mandate to monitor compliance but its practice helps to establish a collective and transparent forum for preventing and addressing controversies. A recent development of regional and national information systems on water and environmental issues under the aegis of the ICWC is one of the examples ...[102]

Of the other agreements considered here, the objectives and principles of the first, the *Saving the Aral Sea Agreement*, are largely found indirectly and in simplified form, in the recitals to the Agreement, which emphasize cooperation and joint management of the Aral Sea Basin, to improve its ecology while maintaining social and economic development. The objective of the Fund is to conclude contracts, have legal status in litigation, and hold and dispose of property 'for the purposes connected to problems of the Aral (Article 2). Indicative of its recent work, in December 2010 the IFAS Board reviewed the draft of the Third Aral Sea Basin Programme and submitted it for approval to the member states. This has been described as: 'a regional action plan for 2011–2015 to alleviate the environmental and socio-economic consequences of the Aral Sea disaster and to facilitate progress towards integrated water resources management (IWRM) and sustainable development in the Aral Sea Basin'.[103] The programme covers four areas, of which the last, 'institutional and legal strengthening of regional water cooperation' is considered 'politically a very difficult area'.[104]

In relation to the second, the *Interstate Commission Agreement*, importantly this confirms the status of the ICWC, and section 1.4 describes it as 'a regional body of the Central Asian states to deal with joint solution of issues related to shared water management, effective use and protection in the Aral Sea basin and to implement ... commonly elaborated programs following the principles of collectivity and mutual respect of parties' interests'. It also confirms that the ICWC is focused upon 'equitable water allocation along the interstate sources, taking into account nature[s] needs and future development' (section 1.5). Comparing international water law to the ICWC however, the latter has been

[102] Dinara Ziganshina, 'International Water Law in Central Asia: Commitments, Compliance and Beyond' (2009) *The Journal of Water Law* 96, 103.

[103] Libert and Lipponen, above n. 40, 569.

[104] Ibid, 570.

found to be deficient, despite lack of objection from the state Parties themselves:

> … the legal system surrounding the ICWC poses contradictions: first, it is excessively sector-specific, focusing on terms of exchange in the trade of water for energy; second, it essentially guarantees the management principles and the exchange structures put in place during the Soviet era, which are structurally unsatisfactory from an environmental viewpoint, but accepted by the region's five states.[105]

The main objectives are found in Part II, which lists 19. They include: 'elaboration and implementation of a regional policy of efficient shared water use and protection' (section 2.1); 'development and implementation of … regional environmental programs' (section 2.5); 'preparation of projects … in area of shared water management' (section 2.8); 'development and operation of unified regional, basin and national information systems on water use' (section 2.10); 'investigation of disputes and disagreements between shared water users' (section 2.17); and 'elaboration of country proposals on construction, reconstruction and operation of interstate water structures' (section 2.19). Dispute resolution, particularly as connected with issues of implementation and compliance is a major weakness of the arrangements for the Aral Sea. In addition to other general legal problems – as indicated below. MacKay emphasizes this as follows:

> A further difficulty has been a lack of provision for dispute resolution between both the new governing bodies and between member states. Disagreements have tended to end in deadlock … Disagreements between states have perhaps been exacerbated by side deals between the states, such as a 1996 agreement between three of the five, excluding Turkmenistan and Tajikistan, over resources trade in the region … A broader difficulty with these legal and administrative projects has been a lack of provision for implementation and enforcement.[106]

Despite the efforts of the central Asian states since independence, it is therefore arguable that the position in the Aral Sea Basin has not improved at all; rather, it has deteriorated significantly. A large part of the environmental problem has been the dam-building activity of the upstream states, such as the Rogun Dam noted earlier, which is a cause

---

105 See Campins Eritja et al., above n. 80, 126.
106 Above n. 88, 23.

of continuing disputes.[107] From a position of Soviet dominance and exploitation to one of subregional competition for resources in accordance with arrangements now recognized under international law, the regulatory response to independence has also failed, as MacKay explains further and forcefully states:

> At the time of the Soviet breakup, there was hope both in the region and elsewhere that regulation under international law would improve the environmental situation. If Soviet regulation had led to environmental ruin, the argument went, then the intervention of international law would help to reverse this – a change in regulatory regime would change the situation. In practice, the result has been rather the opposite – international law has had little practical effect, the absence of overall authority has led to infighting among riparian states, and depletion of the Aral Sea continues ... Where progress has occurred, as in a recent project in Kazakhstan, it has not been the result of regional cooperation or international legal influence.[108]

Such comments are not made lightly. They also have implications for regime-building elsewhere, including the Third Pole, examined in Chapter 5 of this book. Despite the legal measures put in place,[109] governance has ultimately therefore failed. Libert et al. suggest the option of establishing a Commission under the United Nations to solve the problems of the Aral Sea. Although this was proposed by the central Asian Presidents fifteen years ago, it has yet to be tested. The authors believe there are arguments in its favour, notably that 'a reformed structure may be able to coordinate donor activities better'.[110] They also point to other good practice examples which may provide a useful precedent, in particular the experience of the Chu Talas Water Management Commission, which 'is applicable also in other transboundary river basins of Central Asia';[111] this provides for reimbursement by Kazakhstan to Kyrgyzstan for part of the water infrastructure costs relative to the water volumes produced for the former.

Another international initiative which has been reported on by MacKay is a WB grant to Kazakhstan to build a dyke permanently dividing the Aral Sea. This follows support from the WB for an environmental and

---

[107] See above n. 35 and discussion in numerous articles, such as Wegerich, above n. 92, 282-283. Note also the discussion in Libert and Lipponen on the 'safety of hydrotechnical installations' (dam safety), above n. 40, 571, following failures of the Kyzyl-Agash Dam (Kazakhstan) and the Kirov Dam (Kyrgyzstan).

[108] Above n. 88, 17.

[109] See above n. 88.

[110] Libert et al., above n. 86, 19.

[111] Ibid.

social assessment of the Rogun Dam project;[112] such assessments are common in connection with transboundary infrastructure projects, and will be considered in connection with the role of the Chinese-led Asian Infrastructure Investment Bank (AIIB) in Chapter 5.[113] While the dyke would only benefit half of the Sea, MacKay comments:

> … it is cause for hope nonetheless. Results in the northern Aral Sea … were genuinely promising: a substantial rise in sea levels, a reduction in salinity, and increasing fish production … What is striking here is that, despite the involvement of an international organisation, there is little international legal or regulatory influence in play.[114]

This failure of international law is a continuing theme in the Aral Sea subregion, as Libert and Lipponen, subsequently writing in 2012, conclude somewhat dammingly:

> There is currently no fully functional legal framework for the management and protection of shared water resources in Central Asia. The legal framework for cooperation on the Amu Darya and Syr Darya Rivers, put in place in the early 1990s, is increasingly considered to have become outdated. In the past few years, the agreed arrangements on water allocation have not been fully implemented or it has proven impossible to agree on water allocation … The lack of a holistic, rational, equitable, and sustainable approach to the use of transboundary water resources supported by all riparian countries has resulted in tensions and suspicions over water allocation and energy generation, social and economic problems, and environmental degradation.[115]

---

[112] See Wegerich, above n. 92, 283. The latest situation on this is that the Assessment Studies for the proposed Rogun Hydropower Project in Tajikistan have been finalized after the fifth round of riparian consultations was completed and comments from government and civil society stakeholders were carefully considered. See: <http://www.worldbank.org/en/region/eca/brief/rogun-assessment-studies> accessed 29/3/16.

[113] See for example, Simon Marsden, 'Environmental Assessment of Cross-Border Development: China and the Third Pole' (2016) 18 *Journal of Environmental Assessment Policy and Management* 1650009-1–1650009-2.

[114] Above n. 88, 25.

[115] Above n. 40, 565, footnote deleted.

4. THE SUSTAINABLE DEVELOPMENT CONVENTION

4.1 Geographical Context, Historical Background and Institutional Arrangements

Despite the emphasis on water security in accordance with the arrangements above, it is particularly noteworthy that in 2006 the central Asian states concluded a comprehensive subregional environmental agreement, the *Sustainable Development Convention.* The geographical context and historical background is similar to the above arrangements, although this treaty concerns the full breadth of environmental protection issues, of which water is just one. The institutional arrangements of the *Sustainable Development Convention* are located in Article 23, which establishes the COP, and in Article 24, instituting the Secretariat. One of the initial responsibilities of the COP was 'the establishment of other bodies of the Convention as are deemed necessary' (Article 23(9)(a)). Functions of the COP outlined in Article 23(11) also include (f) '[to] establish such subsidiary bodies as may be necessary'; this would enable an NCP to be established. The Secretariat is established by Article 24, consisting of an Executive Secretary and tasked with organizing and conducting COP meetings and related matters. Although the Convention is not in force – and no protocols or subsidiary bodies have been created as a result – its procedures are explained in the next section.

4.2 Objectives, Principles and Procedures, including Implementation and Compliance

The *Sustainable Development Convention* was adopted in 2006 in the city of Ashkhabad (Ashgabat), Turkmenistan, and is also known as the 'Nastoyaschaya Convention'.[116] The Aral Sea is specifically mentioned as a concern of the Parties who 'attach ... great importance to the territory of the Aral Sea Basin' (Article 2(1)); and a general obligation is to 'adopt and implement appropriate measures to improve the ecological situation in the Aral Sea Basin and particularly in the Aral Sea region' (Article 5(6)). It is focused on the sustainable use and development of the resources of the subregion for current and future generations; 'recognising the importance of integrating environmental policy and legislation in the field of environment and development' (Recital 4), and 'legal regulation' (Recital 5). In addition to 'an effort to further develop the

[116] Article 2(1).

existing agreements between the countries of Central Asia' – which would include the *Caspian Sea Convention* – the stated purpose of the *Framework Convention* (Article 3) is therefore:

> … to ensure the effective protection of the environment for sustainable development in Central Asia, including the improvement of ecological environment, rational use of natural resources, as well as [to] reduce and prevent transboundary environmental damage through the harmonization and coordination of environmental policies and actions of the Contracting Parties and by establishing reciprocal rights and responsibilities.

Article 4, entitled 'Principles', sets out that implementation of the objectives is to be in accordance with the UN Charter and principles of international [environmental] law, including: sovereignty over natural resources and the 'no harm' rule; environmental protection being 'an integral part of the process of achieving sustainable development'; the preventative, precautionary and polluter pays principles; rational utilization of natural resources and intra- and inter-generational equity; and cooperation and EIA. Article 4(6) is noteworthy in focusing on the principle that 'cross-border damage and other harmful effects are prevented or reduced through cooperation'; this encompasses EIA and related notification and consultation, and the polluter pays principle.

Article 5 outlines the general obligations, including: cooperation; adoption and implementation of national policies and actions; integration; fair and reasonable use; and the development of protocols. Article 6 concerns environmental protection and sustainable development; there is indirect reference to the use of strategic environmental assessment (SEA) in connection with the development of plans and programmes.[117] Article 6(2)(b) therefore requires Parties to 'ensure that any plans, strategies, projects and activities that may have adverse effects on natural resources and environment in general, were at an early stage of the relevant impact assessment'.

Article 7(1) requires the establishment of monitoring mechanisms, and Article 7(2)–(4) evaluation, which means EIA in both domestic and transboundary applications; Article 7(4) specifically requires Parties to cooperate in the development of a transboundary EIA protocol. Articles 8–13 deal with protection and management of air, water, land, waste, mountain ecosystems and biodiversity, respectively; it is anticipated as

---

[117] Simon Marsden, *Strategic Environmental Assessment in International and European Law: A Practitioner's Guide* (Earthscan/Routledge, Abingdon, 2008).

will be seen for each, that related protocols are developed to provide substance to the obligations in these articles.

Article 8(1) requires Parties to take all measures to limit, reduce and prevent air pollution, whether national or transboundary, while the obligation to develop a protocol establishing rules and procedures for this is found in Article 8(2). Such a protocol may include monitoring structures and sources, a regional mechanism to exchange information on transboundary air pollution, identifying and prioritizing pollutants, and developing and implementing a regional system of indicators.

Article 9(1) requires Parties to take all measures to prevent water pollution and the depletion of water resources and to ensure their sustainable use. The development of protocols to establish rules and commitments for this is also foreseen by Article 9(2). Such protocols are to relate to improving national monitoring structures concerning quality and sources, the establishment of a regional system for transboundary watercourses, identifying and prioritizing pollutants, joint action to ensure drinking water supplies, prevention of downstream pollution by developing appropriate measures, and cooperation concerning sustainable use and protection of transboundary watercourses.

Article 10(1) requires Parties to take measures to 'combat and prevent' land degradation, which is to include desertification. The establishment of protocols is again foreseen in Article 10(2); with rules and procedures that relate to the improvement of national structures, a regional monitoring system, and a mechanism to exchange information (Article 10(3)). Joint action to combat land degradation, and in relation to sustainable agriculture and forestry; sustainable livestock and arid rangeland management is also foreseen (Article 10(3)).

The attention of Article 11(1) is upon waste management, with cooperation required to limit, reduce and prevent pollution from 'industrial and household waste, focusing on pollution by hazardous waste'. The development of a protocol is part of this obligation, providing for national registers of collection places and disposal, including those for transboundary effects; establishing clean production and technologies via a regional network of centres; and the development of measures and action 'to prevent the transboundary spread of radioactive contamination from mining and test sites'.

Article 12 requires Parties to cooperate in taking measures to prevent the degradation of mountain ecosystems, with a protocol establishing rules and procedures for sustainable development of mountain areas; cooperative measures to restore mountain ecosystems which have been disturbed; regional structures for sustainable development generally; and the adoption of joint measures 'to prevent pollution and depletion of snow and ice surfaces' (Article 12(3)). Article 13 relates to biodiversity generally, but also has relevance for the Article 12 requirement given the importance of biodiversity to mountain ecosystems.[118] It obliges Parties to take measures to preserve biodiversity and sustainably use its components (Article 13(1)). In doing so, the development of a protocol is envisaged to improve national systems to identify and monitor its components, establish a regional ecological network, and jointly take measures for on and off site conservation (Article 13(3)).

The requirements in Article 8 – air protection – and Article 12 – mountain biodiversity – in particular take the *Framework Convention* beyond the existing conventions analysed in Chapter 2, with Chapter 3 emphasizing additional concerns to the coastal focus of the Regional Seas agreements, in particular the water management focus of the arrangements for the Aral Sea. The conservation of mountain ecosystems and biodiversity in Articles 12 and 13 is furthermore especially relevant for the protection and management of the Third Pole, which is focused on heavily in Chapter 5 of this book. The Article 12 and 13 provisions are also relevant in the context of the CBD Mountain Biodiversity Programme.[119] A key part of the conservation of mountain ecosystems is however linked with the health of transboundary watercourses, and just as the Regional Seas agreements focus on ICZM, so the focus on IWRM is essential. In the context of central Asia's mountains, this is explained and highlighted as follows:

> [In recent years] a concept of integrated water resources management (IWRM) at national and regional level [has become] ... a basis for [the] rational use of limited water resources of the region. According to IWRM principles, water and associated ... land and other natural resources within [the] catchment area shall be considered as an object of joint use, management, protection and development. Duties and responsibilities shall be shared between the water users at national, sectoral, local and grass-root levels in such way that regulation of water consumption and water use could ensure stable protection and/or development ...[120]

---

[118] See Simon Marsden, 'Asian Precedents for Effective Biodiversity Conservation in the Third Pole', paper presented to *International Symposium: Biodiversity, Sustainable Development and the Law*, Centre for International Sustainable Development Law/Lauterpacht Centre for International Law, St John's College, University of Cambridge, 2015.

[119] <https://www.cbd.int/mountain> accessed 19/5/15.

[120] Interstate Sustainable Development Commission/Regional Mountain Centre for Central Asia, *Strategy of the Regional Mountain Centre for Central Asia*, Bishkek, 2009, 10.

Article 14 is concerned with cooperation in emergencies; Article 15 national authorities; Article 16 scientific and technical cooperation; Article 17 exchange and access to information; and Article 18 public participation. Article 14(1) is a forward-looking provision in connection with emergencies, which it recognizes as both 'man-made and natural [in] origin'. In considering provisions of relevance to the Third Pole, where natural disasters – in particular earthquakes – require an effective response, this is an essential provision. The development of a protocol is once more part of the regulatory scheme, and Parties must cooperate in providing information and monitoring man-made and natural processes leading to emergency situations (Article 14(3)). Under Article 15, Parties must designate a national authority and provide it with the power to implement actions under the Convention. Under Article 16(1), Parties must cooperate both with one another and with third parties to develop capacity, including scientific and technical expertise in relation to environmental protection and management; research efforts must focus on the 'development and deployment of clean technologies and industries' in the context of sustainable development. Research must be shared between Parties (Article 16(2)), who 'shall strive to create and improve regional scientific and technological centres' to take work forward (Article 16(3)).

Further to the concerns expressed regarding access to information in the *Land-Based Sources and Activities Protocol* of the *Caspian Sea Convention* (see above), Article 17(2) requires that: 'In accordance with their national law and taking into account the provisions of existing international treaties relating to public access to environmental information, the Contracting Parties shall strive to ensure public access to information on the status of the environment in Central Asia'. Article 18 concerning public participation is however brief, merely stating that the COP must 'establish rules and procedures for the involvement of civil society in the implementation of the Convention, including by introducing a system of accreditation of NGOs'.

Article 20 contains the NIR requirement that national authorities report to the Secretariat on implementation; Article 19 precedes this with obligations in respect of implementation and compliance. Parties are therefore obliged to 'take such measures as may be necessary', which may include 'adoption or amendment of national legislation' (Article 19(1)); and – in highlighting that compliance failure may be as much due to lack of capacity as deliberate breach – the COP 'shall develop and adopt rules and procedures to promote compliance … including assistance in cases of reasonable compliance' (Article 19(3)). Parties must also cooperate with international organizations, financial institutions and donors 'on issues related to the implementation of this Convention' (Article 19(2)).

Although this treaty is a comprehensive instrument potentially applicable to all of the central Asian states, it has yet to come into force, with little indication of when this may occur. While Turkmenistan, Tajikistan and Kyrgyzstan signed the *Sustainable Development Convention* in 2006, Kazakhstan and Uzbekistan 'continue to consider their participation'.[121] As such, it is not possible to comment on effectiveness in relation to the procedures outlined above. However, central Asia is experiencing particularly rapid change and is being integrated into other parts of Asia at a significant rate, with major potential consequences for the environment and society. In a similar way that gas pipelines are being constructed to export resources from the Caspian Sea to surrounding counties, particularly China, so too is the currently under exploited hydro-electric power (and its potential) drawing attention from scholars, some of whom commented in 2016 as follows:

> The hydroelectric energy sector in Central Asia is not yet fully developed, but it is raising high expectations. Currently, the contribution of hydropower to general energy consumption moves in a very wide range, and it may differ from the ninety eight percent in Tajikistan to the one percent in Turkmenistan. Due to the economic crisis, the internal electricity demand is currently weak, but electricity exports are growing fast. Hydroelectric energy is expected to provide enough electricity to meet the national energy needs of the upstream Central Asian countries at a lower cost, as well as to help avoid energy shortages in the winter season. In addition, surplus power might be exported to other interested actors, such as China and Russia, India and Pakistan, or Afghanistan and other South Asian countries.[122]

5. CHINA'S RELATIONSHIP WITH CENTRAL ASIA

China's relationship with central Asia has many dimensions. Vinokurov and Libman refer to China as a 'new and very successful entrant' in regard to Eurasian integration, as one of the key foreign investors in the subregion.[123] With an increased presence in the energy area in particular, they state that: 'the Chinese government views Central Asia as key to its

[121] Campins Eritja et al., above n. 80, 141.

[122] Campins Eritja et al., above n. 80. 138.

[123] Evgeny Vinokurov and Alexander Libman, *Eurasian Integration: Challenges of Transcontinental Regionalism* (Palgrave Macmillan, Basingstoke, 2012) 217.

policy of diversifying its energy supply sources'.[124] Although China is neither a Party to the subregional agreements outlined above (nor a dialogue partner), this has not prevented its involvement being highly successful. Litta comments: 'Regime theory argues that important actors need to be part of the regime. This has not been proven right. [One of t]he most important countries [is] China … [While it is not a] formal member of agreements resulting from the regime[s] … [it has] obtain[ed] the keys for problem solving'.[125]

The 'New Silk Road' is a major part of China's involvement in central Asia, and this has been key to the discourse of external affairs for at least two decades.[126] The initiative was announced by President Xi in 2013, and is also known as the 'One Belt, One Road' programme. Under this programme China aims to invest primarily in infrastructure projects, including railways and power grids in central, west and southern Asia, as well as in Africa and Europe.[127] The 'One Belt' component is a reference to the terrestrial economic belt, and the 'One Road' a reference to the maritime aspect, which is considered further in Chapter 4 (southeast Asia). The 'Maritime Silk Road' as it is sometimes called, is intended to provide an alternative route in the seas to the south and west, to the degree at least that this is possible.[128] The main routes travelled by this programme have been outlined.[129] The intention is clearly to not only continue China's impressive rates of domestic growth, but also to provide a means of diversifying by exporting development overseas. In turn, development provides China with a way of returning energy to China via power grids and pipelines, and also to ease the passage of domestically produced products to high-value markets abroad, particularly Europe and in rapidly developing southwest and central Asia. The 'New Silk Road'

---

[124] Ibid. See also Jean-Pierre Cabestan, 'Energy Cooperation between China and Central Asia' (European Council on Foreign Relations/Asia Centre, 2011) 6–8.

[125] H Litta, *Regimes in Southeast Asia: An Analysis of Environmental Cooperation* (Verlag Springer, Wiesbaden, 2012) 195.

[126] See Hasan H Karrar, *The New Silk Road Diplomacy: China's Central Asian Foreign Policy since the Cold War* (UBC Press, Vancouver, 2010).

[127] See Peter Wong, 'The Dealmaker', *South China Morning Post*, 30 May 2016.

[128] China's territorial disputes with southeast Asian states is a challenge to the wish of these states to secure Chinese investment and for China to provide it. See Chris Horton, 'Treading Warily on China', *International New York Times*, 26 May 2016.

[129] Minkang Liu, 'Navigating China's New Silk Road', *South China Morning Post*, 18 January 2016.

programme is supported by a considerable fund;[130] this fund was the 'driving force behind the establishment of the Asian Infrastructure Investment Bank', which is analysed in detail in Chapter 5. However, despite this and clear early progress, there have been signs that all is not well given recent domestic economic circumstances in China[131] and the fact that large financial losses may be occurring as a result of the programme.[132]

The collapse of the Soviet Union provided the opportunity for enhanced Chinese engagement with the central Asian republics, as noted by Duran and Yilmaz, although limitations have been acknowledged.[133] The interest of China, alongside many other states with interests in the area, is understandably in the huge natural resources of the subregion that were described in an earlier section. There are however many other dimensions, including the wish to manage the issues of separatism that have beset the westernmost Chinese provinces of Xinjiang and Tibet. Niazi draws attention to the border disputes between China and its central Asian neighbours as a key part both of this and of solidifying relations between them to advance the economic and security imperatives of its strategy.[134] Niazi refers to disputes between Kazakhstan, Tajikistan and Kyrgyzstan in particular; although at the time of writing the Tajikistan dispute relating to the contested Babakhshon region was still contested, and the Kyrgyzstan dispute highly significant because of the shared border with Xinjiang, which in 2006 accounted for 40 per cent of the total trade between China and central Asia. He comments: 'Having

---

130 Mini Chan, 'China to Create US$40 Billion Silk Road Fund to Upgrade Asia Links' *South China Morning Post*, 9 November 2014.

131 See Reuters, 'China's "New Silk Road" Facing Difficulties: Former Official', *South China Morning Post*, 23 April 2016; and Victoria Ruan, 'Silk Road Plan Risks Low Business Return', *South China Morning Post*, 5 January 2015.

132 Ibid. China is therefore attempting to obtain additional support for the programme, including from Hong Kong, above n. 127.

133 Hasan Duran and Kadir Kürşat Yilmaz, 'Energy', in EAH Gencer and C Gerni (eds), *Central Asia Economies in Transition* (Cambridge Scholars, Newcastle, 2012); and Li Hak Yin and Zhengxu Wang, 'China's Advancements in Central Asia: Limitations Toward a Regional Power' (2009) 1(4) *East Asian Policy* 5.

134 Niazi, above n. 31; see also Angela Meng, 'Central Asian Nations to Conduct Terrorism Drills with PLA', *South China Morning Post*, 14 August 2014.

mended its frontiers with its neighbours, China earned a great deal of goodwill, which helped raise its profile in the region'.[135]

Kazakhstan is the subregion's principal oil producer,[136] as well as the largest coal and uranium producer. China has been involved in a number of oil projects with Kazakhstan, the Kashagan oil project of which is notable as the world's largest oil discovery in fifty years.[137] Niazi reports that in 1997 China pledged US$10 billion in investment for oil exploration and infrastructure construction. In 2005 it bought PetroKazakhstan for US$4.18 billion, and also invested US$700 million in pipeline construction works.[138] In 2009 the Kazakh-China oil pipeline doubled its capacity, and China now accounts for around a third of Kazakh oil exports; subsidiaries of the state-owned China National Petroleum Corporation furthermore produce about one fifth of Kazakh oil supplies. Given Kazakhstan is also a transit state, it is perhaps no surprise that China has also been responsible for constructing the network of road and rail links to connect Kazakhstan with Uzbekistan. A recent visit to Kazakhstan by the Chinese President was commented upon by the international media in the following terms:

> The Chinese foreign ministry said the cooperation included the joint construction of the Silk Road Economic Belt proposed by Beijing for trade and infrastructure development along countries in Asia and Europe. Kazakhstan, China's strategic partner for the last 23 years, is one of the countries along the belt. The visit was Xi's second to Kazakhstan in two years, and the trip is expected to increase mutual trust between the two states.[139]

Development of the 'more modest oil fields' in the Fergana Valley of Uzbekistan has also been supported by China, with the first pledge of assistance in 2006 during a visit to China by the Uzbek President. The primary interest of Uzbekistan to China is however in the uranium resources, which are also courted by Russia, having seen its own supplies decline, as Niazi notes.[140] China is also involved in projects in Turkmenistan, which contains the world's fourth largest natural gas reserves.

---

135 Above n. 31, 111.

136 Wojciech Ostrowski, *Politics and Oil in Kazakhstan* (Routledge, London, 2009).

137 <http://www.ncoc.kz/en/kashagan/default.aspx> accessed 12/3/15.

138 Above n. 31, 112.

139 Lawrence Chung, 'Chinese President Xi Jinping Visits Kazakhstan, Russia and Belarus to Discuss Silk Road Belt', *Sunday Morning Post*, 8 May 2015.

140 Above n. 31, 113.

In 2006, a 30-year agreement was signed for the provision of gas to China, with a pipeline also agreed to be completed via Kazakhstan.

However, further to this collaborative oil and gas field development between China and these central Asian states, there has been inadequate attention focused on environmental issues. The construction of the Kazakhstan-China oil pipeline and the Turkmenistan-China gas pipeline are a significant cause for concern in relation to their environmental impacts for example.[141] The China National Petroleum Corporation, as a partner to the first pipeline project, has not carried out any transboundary EIA[142] with respect to this,[143] which, while not required under treaty law, should arguably have been attended to in accordance with customary international law.[144]

Providing a broader context for the relationship between China and central Asia, the establishment of the Shanghai Cooperation Organization has been one means by which China has pursued its successful cooperation with the central Asian states.[145] This cooperation is not limited to interests in oil and gas and other resources, as it provides a market for Chinese produced consumer goods. It has also assisted in connection with its relationship with Russia, with whom – given the past historical relationship between Russia and the central Asian republics – it clearly competes. What remains here is to consider (very briefly) China's relationship with Iran, which – like Russia is also a Party to the *Caspian*

---

141 With respect to the Kashagan project, pipeline leaks have already been detected. See Andrew Callus and Stephen Jewkes, 'Crippled Kashagan Oil Project a Bureaucratic "Nightmare"', *Reuters*, 2 January 2014.

142 Simon Marsden, 'Developing Approaches to Transboundary Environmental Impact Assessment in China: Cooperation through the Greater Tumen Initiative and in the Pearl River Delta Region' (2010) 9(2) *Chinese Journal of International Law* 393–414.

143 See: <http://www.cnpc.com.cn/en/FlowofnaturalgasfromCentralAsia/FlowofnaturalgasfromCentralAsia2.shtml> accessed 12/3/15.

144 See International Court of Justice (ICJ), *Case Concerning Pulp Mills on the River Uruguay (Argentina v Uruguay)*, Judgment, 20 April 2010, paras 204–205, available at: <http://www.icj-cij.org/docket/files/135/15877.pdf> accessed 12/3/15.

145 Thrassy N Marketos, *China's Energy Geopolitics: The Shanghai Cooperation Organization and Central Asia* (Routledge, Abingdon, 2009); Michael Clarke, 'China and the Shanghai Cooperation Organization: The Dynamics of "New Regionalism", "Vassalization", and Geopolitics in Central Asia', in Emilian Kavalski, *The New Central Asia: The Regional Impact of International Actors* (World Scientific, Singapore, 2010) 117–148; and Andrei Kazantsev, 'Russian Policy in Central Asia and Caspian Sea Region' (2008) 20(6) *Europe-Asia Studies* 1073, 1078–1079.

Sea Convention.[146] Iran is the largest state in the central Asian subregion, and it has had a developing relationship with China in recent years.[147] A report in the international press, noted the various aspects of this in the following terms:

> China helped mediate the Iranian nuclear deal last July and Xi will be the first state leader to visit Iran since the nuclear-related sanctions were lifted on January 17.
>
> The mainland has been Iran's biggest trade partner for the past six years and the biggest buyer of its crude oil and non-petroleum products.
>
> Trade volume between the two nations surpassed US$50 billion in 2014 – up 31.5 per cent on the volume of trade the previous year.
>
> In 2014 China's export volumes to Iran in 2014 totalled US$27.5 billion, while the volume of imports from Iran totalled US$24.3 billion.
>
> The mainland exports machines and electronics, textiles, chemical and steel products and import[s] crude oil, ore and agricultural products.[148]

As with its relationship with Saudi Arabia as considered in the previous chapter, ensuring the environment of the subregion is adequately protected will be a key test of the *Caspian Sea Convention.* How China uses the 'keys for problem solving' that were noted by Litta above will no doubt be watched closely to see if the environmental protection objectives of the Convention can be reconciled with much greater demand for resource exploitation.

6. CONCLUSIONS

The environmental governance of central Asia is, to put it very simply, very complicated. It is complicated by the history of Soviet occupation and the interests of numerous states in the subregion – Russia, as successor to the Soviet Union, China as emerging superpower, and the traditional interests of Iran and the USA among others. Its natural

---

[146] Russia's relationship with central Asia is beyond the scope of this book; see however Kandiyoti, above n. 5; Vinokurov and Libman, above n. 26; and numerous other references above.

[147] For consideration of Iran's relationship with the central Asian states, see Kandiyoti, 'Iran, the idle hinge' section of article, above n. 5, 86.

[148] Pinghui Zhuang, 'Key Facts Behind China's Warming Ties with Saudi Arabia, Iran and Egypt as Xi Jinping Signs Mega Oil Deals During his Middle East Tour', *South China Morning Post*, 20 January 2016.

resources in oil, gas and uranium in particular make it of interest to numerous states, multinational corporations and lending banks. Its location – adjacent to Eurasia, south Asia and east Asia (and north Asia if Russia is so termed) – adds to the complexities because of the strategic significance in relation to each of these subregions and regions. Whether as a strategic support for the west in its relationship with, for example, Afghanistan, or for the east, with China attempting to manage and secure its territorial integrity in the face of conflict in Xinjiang, it is a sub- or cross-subregion/region of great importance.

The *Caspian Sea Convention* is, like the *Red Sea/Gulf of Aden Convention* analysed in Chapter 2, therefore far more than a subregional regime. The geographical location of the Caspian Littoral States demonstrates that it is also a cross-region and cross-subregion, given the intersections between Europe and Asia (Russia in particular), and Iran (southwest Asia), Kazakhstan and Turkmenistan (central Asia); Azerbaijan's membership (Caucasus, Eurasia and western Asia) adds to this. Furthermore, the involvement of Russia, the world's largest nation geographically, with a population of 143.5 million people, together with Iran, with a population of 77.5 million, and Kazakhstan, by far the largest central Asia state, with a population of 17 million, adds to this significance, especially given the superpower status of Russia and Iran's gradual resumption of global citizenship following its nuclear isolation.

The consequences of oil and gas development in particular have been highlighted in this chapter as of the greatest importance. The growing demands on these resources from Russia, China and other states highlights the need for environmental safeguards to be in place to manage the negative consequences of resource exploitation, which given past events in both the Caspian Sea and globally – notably blowouts from wells, and leaks from pipelines – are not insignificant. The environmental regime for the Caspian Sea, comprising a treaty and four protocols, is an important part of these environmental safeguards. Both the treaty, and in particular the protocols, contain detailed provisions for environmental protection of the Caspian Sea. However, limited ratification of the protocols, with none in force, is a cause for considerable concern given the pace of development and failure of this regulation to keep pace. In addition, a close read of the provisions in the agreements reveals that certain requirements lack the prescription required, for example regarding invasive alien species, as noted by one scholar.[149] While national implementation of certain obligations has taken place, in Iran for

149 Janusz, above n. 55.

example, as identified by others,[150] more is apparently needed. Comparisons with the Regional Seas Agreements in southwest Asia, as analysed in Chapter 2, also identify areas needing improvement.

Turning to the arrangements for the Aral Sea, again the complexities of history and location are at the forefront, and continue to influence the effectiveness of the measures put in place over the last twenty-five years. The distinction between the hydro-electricity needs of the upstream states (Kyrgyzstan and Tajikistan) on the one hand, and the irrigation needs of the downstream states (Turkmenistan, Kazakhstan and Uzbekistan) on the other, reflects this history and location. The movement from a previous position whereby the domestic water management rules favoured the needs of the downstream, generally more dominant states, to one whereby the international rules that replaced them have, while supposedly maintaining the previous principles of water allocation, ultimately not been supportive of either group of states, illustrates the confusion and difficulty of resolving differences for the two rivers that feed into the Aral Sea.

The biggest loser has ultimately been the Aral Sea itself, which despite the urgency of the situation reflected in the naming of the *Saving the Aral Sea Agreement*, has ultimately not been saved. Rather, it has shrunk, been divided and become more saline – a testament to the failed environmental governance of the Basin. The *Interstate Commission Agreement* has similarly been found wanting, with the ICWC unable to apply an intersectoral approach to cooperation. The absence of a compliance control system, which has furthermore not been mandated for the ICWC, is compounded in turn by the failure to provide for any dispute resolution provision. The reason, as noted by one author in connection with the 'institutional and legal strengthening of regional water cooperation', is clear, as it is considered 'politically a very difficult area'.[151]

This is the case, in part, not only because of the lack of political will demonstrated by any of the Parties, but also because of the contradictory nature of the legal system: sector-specific on one hand, and guaranteeing the continuance of inadequate environmental management principles from the past. Ultimately, the international law that has replaced the previous domestic law of the Soviet Union has failed completely to improve the environmental situation, with non-legal contributions, such as multilateral financial support for the dyke construction from the World Bank, considered of greater value.

---

150 Pak and Manuchehr, above n. 3.

151 Libert and Lipponen, above n. 40, 569.

The *Sustainable Development Convention* was the final environmental regime considered in this chapter, which as noted, is not in force; neither have any protocols or subsidiary bodies been established. It is not therefore possible to say anything about effectiveness in connection with implementation and compliance, and the brief concluding comments here are limited to the potential of its provisions. In the light of the remarks concerning the Caspian Sea, and particularly Aral Sea governance arrangements, it is fair to say that the *Sustainable Development Convention*, if detailed with protocols and adherence to the provisions within, could however go some way to respond to the environmental crisis in the Aral Sea in particular. Certainly the objective to 'adopt and implement appropriate measures to improve the ecological situation in the Aral Sea Basin', appears overwhelmingly in need of support. The inclusion of a detailed suite of environmental principles is additionally very supportive of these aspirations, and detailed requirements to 'combat and prevent' land degradation, for example, are clearly needed. Furthermore, obligations for monitoring, NIR and compliance control would, if brought into effect, go some considerable way to seeing the objective achieved.

Finally, in connection with China's role in central Asia, this may either help or hinder efforts for environmental protection, depending on the safeguards that are in place for resource development and exploitation. Accepting that the environment and resources will not be left undeveloped, the need is there for this to be achieved sustainably. The establishment of the New Silk Road programme of infrastructure support, added to the oil and gas development, does not come without risk and environmental consequences. Pipeline leaks in connection with the Kashagan project and failure to conduct general international law obligations of transboundary EIA, are certainly not supportive of confidence building efforts, and more must be done to show that transboundary cooperation, and the avoidance of harm, are not hostage to national gain.

4. Southeast Asia: the Mekong, conservation and haze pollution regimes

1. INTRODUCTION

Moving eastward from southwest and central Asia, this chapter analyses the governance arrangements for three regimes in southeast Asia, the first concerned with a major transboundary river basin, the second measures for conservation, and the third an air pollution treaty. After ending this introduction by defining the subregion of southeast Asia and the main institutional body responsible for governance, the first main section will therefore evaluate the *Agreement on the Cooperation for the Sustainable Development of the Mekong River Basin* (the *Mekong Agreement*)[1] and related measures. This is considered first because its history precedes that of the other regimes, and since the focus of the regime – a major watercourse – is highly significant in the context of the proposed governance of the Third Pole in Chapter 5. It also relates well to the measures for water management in the Aral Sea subregion analysed in Chapter 3, and notably the conflict between upstream and downstream users over hydro-electricity on one hand, and agricultural irrigation on the other.

The second and third sections will examine the two Association of Southeast Asian Nations (ASEAN) environmental regimes. As noted in this introduction below, ASEAN is the primary body through which governance of southeast Asia is managed, notwithstanding the specific institutional role accorded to the Mekong River Commission (MRC). The first (and second regime) is constituted by the related *Agreement on the Conservation of Nature and Natural Resources* (the *Conservation Agreement*),[2] and the *Agreement on the Establishment of the ASEAN Centre for*

[1] Agreement on the Cooperation for the Sustainable Development of the Mekong Basin (Chiang Rai, 5 April 1995) 2069 UNTS 3, in force 5 April 1995.

[2] Agreement on the Conservation of Nature and Natural Resources (Kuala Lumpur, 9 July 1985), not in force.

Biodiversity (*Centre for Biodiversity Agreement*).[3] The second (and third regime) is the *Agreement on Transboundary Haze Pollution* (the *Haze Pollution Agreement*).[4] Although the background and role of ASEAN will precede the outline of all three regimes given its dominant role in southeast Asia today, it will also be considered extensively in connection with the second and third regimes in this chapter. As in Chapters 2 and 3, once more subsections will outline the geographical context, historical background and institutional arrangements; objectives, principles and an overview of procedures including matters of implementation and compliance; and, for the treaties in force, evidence of practice. Again as in Chapters 2 and 3, a penultimate section will analyse China's role, before conclusions follow.

1.1 Defining Southeast Asia and the Significance of ASEAN

Under the UN Geoscheme, 'southeastern' Asia consists of Vietnam, Cambodia, Laos, Thailand, Indonesia, Malaysia, Singapore, Brunei, Myanmar, the Philippines and Timor Leste.[5] As with the subregions of southwest and central Asia, there is significant economic and political diversity between them, although this is even more the case with respect to southeast Asia, especially in terms of religion and history.[6] The subregion shares with southwest Asia a history of varied colonial occupation, but unlike southwest Asia is religiously far more diverse. In population terms, it is also much larger than southwest or central Asia, with around 600 million people. Indonesia has the largest population (250 million), followed by the Philippines (100 million), Vietnam (92 million) and Thailand (65 million). Unlike both southwest and central Asia – and because of its location in the tropics with heavy rainfall – the subregion has abundant freshwater resources and extensive forests covering almost half of the land area. As a result, both have been heavily exploited for electricity generation and timber. Furthermore, an increasing population combined with rapid economic growth has exerted significant pressures, resulting in numerous national and transboundary

---

[3] Agreement on the Establishment of the ASEAN Centre for Biodiversity (Capitals of each member state, 9 August 2005), in force 8 July 2009.

[4] Agreement on Transboundary Haze Pollution (Kuala Lumpur, 10 June 2002), in force 25 November 2003.

[5] <http://unstats.un.org/unsd/methods/m49/m49regin.htm#asia> accessed 11/1/16.

[6] Gavin W Jones, *The Population of Southeast Asia* (Asia Research Institute, Working Paper Series 196, National University of Singapore, 2013).

environmental issues, such as air, water and land pollution, and depletion of biological diversity. Contreras adds:

> Southeast Asia also plays an important role in global environmental security discussions. This is due to its high biodiversity levels, as well as its high rate of species extinction. For example, 59.6 percent of Indonesia's 29,375 vascular plant species are endemic and not found anywhere else. Southeast Asia encompasses 4 of the 20 'biodiversity hotspots' identified in the world. Its rate of deforestation is the highest of any major tropical region, and it could lose 75 percent of its original forest cover and 42 percent of its biodiversity by 2100.[7]

ASEAN is the primary intergovernmental forum in relation to environmental (and other) issues in southeast Asia, consisting of all the nations above except Timor Leste. As it has been extensively discussed elsewhere, this section will provide only a very brief snapshot of relevant institutional arrangements.[8] ASEAN has been involved with environmental governance for a considerable time – particularly since 1997 – although it has been said that 'there has been an admittedly slow and often uneven trajectory of institutionalization of environmental governance arrangements'.[9] An ASEAN Ministerial Meeting on Environment (AMME) is held periodically, '[r]ecognising the importance of environmental cooperation for sustainable development and regional integration …'.[10] Cooperation has recently focused on ten priority areas reflected in the Blueprint for the ASEAN Socio-Cultural Community 2009–2015; the first of these is

---

[7] Antonio P Contreras, 'Transboundary Environmental Governance in Southeast Asia', in A Pandya and E Laipson (eds), *Transnational Trends: Middle Eastern and Asian Views* (The Henry L Stimson Center, 2008) 155, 155.

[8] See for example, K Kheng-Lian and MS Karim, 'South East Asian Environmental Legal Governance' in S Alam, HJH Bhuiyan, TMR Chowdhury and EJ Techera, *Routledge Handbook of International Environmental Law* (Routledge, Abingdon, 2013) 461; Kheng-Lian Koh and Nicholas A Robinson, 'Regional Environmental Governance: Examining the Association of Southeast Asian Nations (ASEAN) Model', in Daniel C Esty and Maria H Ivanova (eds), *Global Environmental Governance: Options and Opportunities* (Yale School of Forestry and Environmental Studies, 2002) 101–120; and B Boer, R Ramsar and DR Rothwell, *Environmental Law in the Asia Pacific* (Kluwer, London, 1998), which includes chapters on both ASEAN (ch 12) and the Mekong (ch 10).

[9] Lorraine Elliott, 'ASEAN and Environmental Governance: Rethinking Networked Regionalism in Southeast Asia' (2011) 14 *Procedia – Social and Behavioural Sciences* 61.

[10] <http://environment.asean.org/about-us-2/> accessed 11/1/16.

addressing global environmental issues, and the second transboundary environmental pollution.[11]

The significance of ASEAN to southeast Asia has, significantly, also resulted in considerable legal and policy scholarship analysing its effectiveness.[12] Improved implementation was noted over a decade ago by Koh and Robinson as the key to this, together with an ecosystem approach, with reference to both the Mekong River and haze pollution, that are the focus of two of the three remaining parts of this chapter. They comment: '... more effective systems for environmental governance need to define active implementation procedures ... ; ASEAN's collaborative systems of regional environmental governance ... need to function on an ecosystem wide basis'.[13] The haze pollution problem has tested the abilities of ASEAN to the full, and the fact that this remains a major issue over the fifteen years since it was identified as 'a real test for ASEAN', suggests – as will be seen below – that it has largely failed on this count.[14] Reporting generally more recently on ASEAN's environmental governance, Elliott comments unequivocally:

> The influence of the behavioural and procedural norms of the ASEAN way has meant that for much of its history, environmental cooperation in Southeast Asia has been confined to functional cooperation and characterized by a weak form of institutionalism and a reliance on national institutions rather than some form of centralized, regional bureaucracy. This reflects the ASEAN preference for non-binding agreements and a general reluctance to interfere

[11] Ibid. See also Bangkok Resolution on ASEAN Environmental Cooperation (Bangkok, 26 September 2012). For a comprehensive collection of the agreements produced, see: Koh Kheng-Lian, *ASEAN Environmental Law, Policy and Governance Selected Documents* (Volumes I and II) (World Scientific, Singapore, 2013).

[12] For early reviews, see: Wakana Takahashi, 'Environmental Cooperation in Southeast Asia (ASEAN)', in Kazu Kato and Wakana Takahashi (eds), *Regional/Subregional Environmental Cooperation in Asia* (Institute for Global Environmental Strategies, 2001) 31–50; Koh Kheng-Lian and Nicholas A Robinson, 'Strengthening Sustainable Development in Regional Governmental Governance: Lessons from the "ASEAN Way"' (2002) 6 *Singapore Journal of International and Comparative Law* 640–682; and Nathan Badenoch, *Transboundary Environmental Governance: Principles and Practice in Mainland Southeast Asia* (World Resources Institute, 2002).

[13] Kheng-Lian and Robinson, ibid, 679.

[14] Takahashi, above n. 12, 48.

in, or to direct in any authoritative way, the environmental practices of member states.[15]

In 2007, on the Fortieth Anniversary of ASEAN, the ASEAN Charter was concluded. It was a significant milestone in formalizing current institutional arrangements, as it established the legal and institutional framework for ASEAN, with Article 3 conferring legal personality upon the institution.[16] Article 1(9) also includes the purpose: 'To promote sustainable development so as to ensure the protection of the region's environment, the sustainability of its natural resources, the preservation of its cultural heritage and the high quality of life of its peoples'.[17] Institutionally, the Charter makes provision for the ASEAN Summit, the key policy-making body comprising Heads of State, which is to be held twice annually (Article 7); and the ASEAN Coordinating Council, comprising Foreign Ministers, which is to also meet twice a year, and which is responsible for the coordination of the implementation of the ASEAN agreements (Article 8).

Ministerial bodies are established under Article 10, which includes those involved in the Ministerial Meeting on the Environment. The Secretary General, the chief administrative officer, is also responsible for monitoring the implementation of agreements (Article 11). In accordance with the historical underpinnings of the 'ASEAN Way', referred to by Elliott above, the decision-making process continues to be based upon consultation and consensus (Article 12); disputes are similarly to be determined ultimately by the ASEAN Summit (Article 26), including issues of non-compliance (Article 27); however recourse to multilateral dispute settlement is also available in some circumstances (Article 28). Importantly, Article 41 makes provision for external relations with other countries or organizations; Article 44 establishes the status of external parties such as Dialogue Partners and Special Observers; Article 45 relations with other international bodies, including the United Nations; and Article 46 accreditation of non-ASEAN member states.

The current ASEAN Institutional Framework for Environmental Cooperation connects the Summit, Coordinating Council and related

---

[15] Lorraine Elliott, 'East Asia and Sub-Regional Diversity: Initiatives, Institutions and Identity', in L Elliott and S Breslin (eds), *Comparative Environmental Regionalism* (Routledge, Abingdon, 2011) 56, 60.

[16] Charter of the Association of Southeast Asian Nations (Singapore, 20 November, 2007).

[17] Ibid, Article 1(9).

environmental bodies.[18] It also includes a number of ASEAN Working Groups responsible for environmental protection. ASEAN is relevant in this chapter for many reasons therefore, not only for the fact that it has developed and is responsible for the agreements analysed in the final part. In relation to the Mekong River Basin, which is considered first in the next part, it has also had a major role. Litta highlights that this can be traced back to before Laos, Cambodia and Vietnam joined ASEAN, when it linked 'the non-Mekong and Mekong members of ASEAN together for the first time in order to enhance cooperation in the region and also to foster peace and stability in the Indochina Peninsula'.[19] In 1995, the same year the *Mekong Agreement* was adopted, it declared its Mekong Basin Development Cooperation, and proposed two railway projects linking major cities on the peninsula. It has collaborated strongly with the Asian Development Bank (ADB) in development activity since, with the shared goal of poverty reduction. ASEAN has in fact cooperated with many bodies in the subregion, as noted by Elliott:

> ASEAN has never stood alone in its regional efforts. Southeast Asian governments participate in environment, sustainable development, and natural resource management programs and projects under the auspices of a number of intergovernmental supra-regional bodies. The most prominent are the ADB, UNESCAP, and the regional Asia Pacific offices of UNEP and the United Nations Development Programme. Each has, in turn, spawned a range of initiatives and agreements focused wholly or in part on Southeast Asia, some intended to support the efforts of individual governments and some directed more to enhancing regional responses.[20]

2. THE MEKONG AGREEMENT

2.1 Geographical Context, Historical Background and Institutional Arrangements

The Mekong River Basin is both a unique ecosystem and a source of livelihood for a significant majority of the Basin's 65 million people. The Basin is one of the most biodiverse areas on earth, and since 1997 over 1500 new species have been found. Per unit area, the Mekong has more

---

18 <http://environment.asean.org/about-us-2/> accessed 11/1/16.

19 H Litta, 'Case Study Two: Common Use of the Mekong', in *Regimes in Southeast Asia: An Analysis of Environmental Cooperation* (Verlag Springer, Wiesbaden, 2012) 139, 162.

20 Elliott, above n. 15, 65.

fish species than the Amazon, over 1200 in total; most are migratory, and the fishery is the most productive of all inland fisheries globally. This is significant for the livelihoods of those who live in the Basin, as is agriculture, which accounts for over 60 per cent of incomes. The source of the River lies in the Tibetan Plateau, and the watercourse flows for almost five thousand kilometres from there to the South China Sea. Floods during the wet season account for almost 90 per cent of the total river flow from May to November each year. The consequent flooding events are crucial to this as they carry and spread silt upon the floodplains and coastal habitats where crops are grown. The Mekong Delta is where the river meets the sea, and it is highly dependent upon these flooding events for its health.[21]

The Basin is divided into Upper and Lower parts. The source originates in the Upper Mekong Basin, flowing through China's Yunnan Province. The territory of the five riparian states makes up the Lower Mekong Basin. Thailand is the wealthiest of these states and Cambodia the poorest. Tongle Sap Lake and its associated water system comprises the largest freshwater body in southeast Asia and this is located in Cambodia. Together with Laos, the next poorest of the riparian states, the territories of these states make up the largest part of the Mekong Delta, and almost all of their territory falls within the Basin. Alongside Thailand, Vietnam is the other regional power within the Basin. Burma the remaining riparian, has only a short section of its border with the River, and has played a very limited role in the regime. China however has a major role as the upstream riparian, as will be examined in the penultimate section of this chapter. There are similarities with the states in the Aral Sea when it comes to these dynamics as well as the environmental issues to be addressed by transboundary governance. Ponte comments on the relationship between the Mekong states as follows:

> In the geopolitical context of the Greater Mekong, the dichotomy upstream/downstream is a valuable explanatory characteristic of the riparian's preferences and interests within the river regime. The increasing development and expansion of navigation and hydropower sectors of the Mekong regime, and the damming and re-routing of the river these implicate, promulgate this asymmetric relationship and set up a foundation for the distribution of power

---

[21] Fabrice G Renaud and Claudia Kuenzer (eds), *The Mekong Delta System: Interdisciplinary Analyses of a River Delta* (Springer, Dordrecht, 2012). In relation to climate change, see Mike Ives, 'Climate Change Could Worsen Mekong Delta Woes', in Isabel Hilton (ed), *Forging a New Course for the Mekong* (China Dialogue) 41–43.

> amongst the riparian countries. I identify China, Burma, Thailand and Lao PDR as upstream countries and Cambodia and Vietnam as downstream countries.[22]

International institutional support for water resource management and development of the Mekong can be traced back to 1947, and the work begun at that time by the UN Economic Commission for Asia and the Far East (UNECAFE); this was subsequently to become known as the UN Economic and Social Commission for Asia and the Pacific (UNESCAP) in 1974.[23] In 1951 UNECAFE initiated a technical paper, published the following year, to examine the flood patterns of the Lower Mekong Basin, which were to inform development over the following forty years and 'sets the first combined initiative towards a coherent policy for water management'.[24] In October 1957 the Committee for Coordination and Investigations of the Lower Mekong Basin ('the Mekong Committee') was established, 'driven by foreign funds and experts'.[25] The major tangible outcome in the years following was the completion of the Nam Ngum Dam, which linked Laos to Thailand through electricity sales. The political environment made progress difficult; from the mid-1960s to the early 1970s, the Vietnam War had a major impact:

> … the Mekong became embroiled in Cold War geopolitics as a symbol of cooperation. Cooperation among the four basin states was constantly invoked as a means of bringing peaceful political relations to the region … The United States had championed cooperative development of the Mekong since the creation of the Mekong Committee, and officials and engineers from both the United States Bureau of Reclamation and Army Corps of Engineers played prominent roles in the Committee's early planning and hydrologic activities. In addition, the US government's promotion of Mekong cooperation was a key part of its more general geopolitical strategy to funnel aid dollars towards economic development projects in the region, particularly Thailand, as ideological and material counterpoints to socialism.[26]

---

[22] Elena Ponte, 'Preliminary Analysis of Water Security in the Mekong River Basin' UNU-CRIS Working Papers W-2012/12 (United Nations University, Brugge, 2012) 1, 5.

[23] <http://www.unescap.org/about> accessed 30/3/16.

[24] Above n. 22, 14.

[25] Above n. 22, 15.

[26] Chris Sneddon and Coleen Fox, 'Rethinking Transboundary Waters: A Critical Hydropolitics of the Mekong Basin' (2006) 25 *Political Geography* 181, 185.

US support for Thailand continued through the 1980s, and this support explains in part its strong position today. However, developments in the Mekong Basin slowed during the civil war in Cambodia, with a decrease in plans for the Basin from 180 to 29 from the *Indicative Basin Plan* of 1970 to the *Revised Indicative Basin Plan* of 1987.[27] It was only at the end of conflict in 1991 that efforts resumed, with 'the governments of the region once again contemplat[ing] the Mekong's resources as a source of rapid economic development due to its vast hydro-electric potential and capacity to store water for irrigation schemes'.[28] Support from international organizations such as the UN, WB, ADB and bilateral aid agencies provided assistance to these efforts. The ADB has been particularly significant, with its Greater Mekong Subregion (GMS) Programme regarded as 'the principal framework for directing economic development assistance'.[29] Dore emphasizes the role of the ADB, explaining that from early beginnings in 1992 and focused on economic cooperation, it subsequently 'had to review its approach to "environment" – ecological and social dimensions – in response to a sharpened social and environmental ethic, criticisms of past Bank approaches, the demands of shareholders and research advocating alternatives'.[30]

In April 1995, Cambodia, Laos, Thailand and Vietnam signed the *Mekong Agreement*. The event was the culmination of many years of collaboration, from the acknowledged founding of the Mekong Committee almost 40 years earlier,[31] '... the largest single development project the fledgling United Nations organisation had ever undertaken'.[32] Article 36B of the *Mekong Agreement* replaces the previous arrangements.[33] The Mekong was one of the world's great untamed rivers and there was a renewed wish to capitalize on its economic potential. No international river body had ever attempted to assume such responsibilities. Hundreds

27 Ponte, above n. 22, 15.

28 Sneddon and Fox, above n. 26, 187.

29 Ponte, above n. 22, 16.

30 John Dore, 'The Governance of Increasing Mekong Regionalism', paper presented to Regional Centre for Social and Sustainable Development conference, Chiang Mai University, 11–14 July 2003, 12.

31 Recital 1, Preamble, Mekong Agreement.

32 See <http://www.mrcmekong.org/about-mrc/history/> accessed 12/1/16.

33 The Statute of the Committee for Coordination of Investigations of the Lower Mekong Basin of 1957 as amended, the Joint Declaration of Principles for Utilization of the Waters of the Lower Mekong Basin of 1975, and the Declaration Concerning the Interim Committee for Coordination of Investigations of the Lower Mekong Basin of 1978, and all Rules of Procedures adopted under such agreements.

of surveys were undertaken during this time, and the successor of the Mekong Committee, the MRC created by the *Mekong Agreement*, represented a 'coming-of-age' for the original river basin agency, being no longer under the umbrella of other organizations. Thenceforth, the management responsibility lay in the hands of its four members.

The sustainable development of the Basin was – and is – however a major challenge. All six countries in the region, the four state Parties, known as 'Member Countries' (Cambodia, Laos, Thailand and Vietnam) to the *Mekong Agreement*, and the two 'Dialogue Partners' involved in the cooperative framework (China and Myanmar), are riparian countries which depend upon the River as a source of economic development. Plans for dam building have been ongoing for some time, and those completed have led to significant environmental impacts on the ecosystem. Impacts include a reduction in the flooding events, the silt from which farmers depend upon for rice production; floods also reduce water salinity that comes from the river delta.[34] Impacts upon inland fisheries, and on the other aquatic life upon which they depend, are also a problem.[35] Furthermore, in more recent decades, the effects of climate change on water flows have been acknowledged for their significance. In the 2014 Intergovernmental Panel on Climate Change Report transboundary pollution is also identified as an intra- and inter-regional issue of significance.[36]

The main institution created by the Agreement is therefore the MRC. This is established by Article 11 of the Agreement and is given international legal personality to enter into agreements and other obligations with global bodies. The structure of the MRC is confirmed in Article 12 to consist of three bodies: the Joint Committee, Council and Secretariat, with the first of these to perform the substantive functions in relation to the Agreement outlined in the next paragraph. With respect to the Council, its composition – one member of each State at the

---

34 Badenoch, above n. 12, 3; Sneddon and Fox, above n. 26, 188–189.

35 Sneddon and Fox, above n. 26.

36 See Y Hijioka, E Lin, JJ Pereira, RT Corlett, X Cui, GE Insarov, RD Lasco, E Lindgren and A Surjan, 'Asia', in VR Barros, CB Field, DJ Dokken, MD Mastrandrea, KJ Mach, TE Bilir, M Chatterjee, KL Ebi, YO Estrada, RC Genova, B Girma, ES Kissel, AN Levy, S MacCracken, PR Mastrandrea and LL White (eds), *Climate Change 2014: Impacts, Adaptation, and Vulnerability. Part B: Regional Aspects. Contribution of Working Group II to the Fifth Assessment Report of the Intergovernmental Panel on Climate Change* (Cambridge University Press, Cambridge, 2014) 1327, 1355 <http://ipcc.ch/pdf/assessment-report/ar5/wg2/WGIIAR5-Chap24_FINAL.pdf> accessed 12/1/16.

Ministerial and Cabinet level – is outlined in Article 15; the Chair is to rotate on an annual basis (Article 16), and sessions are to be at least every year (Article 17). Three Council functions are found in Article 18. The first (Article 18A) is to make policies and decisions and provide guidance in relation to the Basin; the second (Article 18B), includes approval of Joint Committee Rules and the basin development plan; and the third is to deal with any disputes arising. With regard to the Secretariat, the technical and administrative body (Article 28), its functions are set out in Article 30(A–G), essentially carrying out decisions and tasks allocated by the Council and Joint Committee.

In regard to the Joint Committee, Article 21 confirms that it is comprised of one member from each State, 'at no less than Head of Department level'. Like the Council, the Chair is to rotate on an annual basis (Article 22), and at least two regular sessions are to be held every year, with potential to convene special sessions and invite observers as appropriate (Article 23). Key functions carried out include: implementation of policies and decisions of the Council and related tasks (Article 24A); Basin planning (Article 24B); collation and exchange of information (Article 24C); and carrying out studies and assessments 'for the protection of the environment and maintenance of the ecological balance' of the Basin (Article 24D). They also include assigning tasks to the Secretariat (Article 24E); dealing with issues and differences between Council sessions (Article 24F); reviewing and approving studies and training (Article 24G); and making recommendations to the Council relating to the structure of the Secretariat (Article 24H). The Rules of Procedure of the Joint Committee are determined according to Article 25, and the Rules for Water Utilization and Inter-Basin Diversions (see section below) according to Article 26.

Further to the carrying out of studies and assessment (Article 24D), there have been numerous reports produced in recent years. Notable among these in the years 2010, 2012 and 2014, were reports examining: the strategic environmental assessment (SEA) of hydropower on the Mekong mainstream;[37] water, energy and food security in connection

[37] International Center for Environmental Management, *MRC Strategic Environmental Assessment of Hydropower on the Mekong Mainstream: Final Report* (Mekong River Commission, Hanoi, 2010); and International Center for Environmental Management, *MRC Strategic Environmental Assessment (SEA) of Hydropower on the Mekong Mainstream: Summary of the Final Report* (Mekong River Commission, Hanoi, 2010).

with transboundary river basin management;[38] and cooperation for water, energy and food security under climate change.[39] The first of these reports – SEA of hydropower – is considered below in connection with the Procedures for Notification, Prior Consultation and Agreement. In relation to the 2012 report, this followed the Mekong2Rio Conference held in May 2012 in Thailand and was a contribution to, and part of the preparations for, the subsequent Rio+20 UN Conference on Sustainable Development.[40] It highlighted the critical connections between water, energy and food security, recommending they be addressed in an integrated way by close collaboration of all actors involved. This 'nexus approach' was explained as follows:

> The nexus approach acknowledges the links between water, energy and food in management, analysis, planning and implementation. In doing so, water related strategies and plans are designed in collaboration between relevant authorities and stakeholders, with the aim of avoiding cross-sector impacts and, perhaps more importantly, looking for combined solutions and synergies for more efficient resource use. The trade-offs that may arise are analysed and discussed considering the relevant water, energy and food security issues as well as potential impacts on or relations to environment, climate, people's livelihoods and other economic sectors. In a transboundary context, additional aspects arise to do with identifying solutions and synergies across state boundaries and analysing and agreeing on trade-offs for decision making.[41]

In relation to the 2014 report, this resulted from a MRC conference held in Ho Chi Minh City 2–3 April 2014. It was organized in advance of the Second Summit of Prime Ministers of the MRC Member Countries to share and learn from experiences globally that may have implications for the Mekong Basin. The global UN Sustainable Development Goals (SDGs)[42] were expected to provide the operational framework for

---

[38] Hanne Bach, Jeremy Bird, Torkil Jønch Clausen, Kurt Mørck Jensen, Rane Baadsgaard Lange, Robin Taylor, Vitoon Viriyasakultorn and Aaron Wolf, *Transboundary River Basin Management: Addressing Water, Energy and Food Security* (Mekong River Commission, Vientianne, 2012).

[39] Hanne Bach, Paul Glennie, Robin Taylor, Torkil Clausen, Fritz Holzwarth, Kurt Mørck Jensen, Abel Meija and Susanne Schmeier, *Cooperation for Water, Energy, and Food Security in Transboundary Basins under Changing Climate* (Mekong River Commission, Vientiane, 2014). See also Hijioka et al., n. 36 above.

[40] <https://sustainabledevelopment.un.org/rio20> accessed 14/11/16.

[41] Above n. 38, 13.

[42] <http://www.un.org/sustainabledevelopment/sustainable-development-goals/> accessed 1/4/16.

cross-sectoral implementation of the findings of the conference, although it was acknowledged that these have a national focus 'which makes it difficult to include transboundary aspects …'.[43] Differences between the focus in 2014 with that in 2012 were highlighted as follows:

> This [2014] conference … demonstrated that transboundary basin management discussions are moving away from a water centred approach towards an integrated approach with all water-related sectors. This is evident when comparing the report from the 'Mekong2Rio' conference held in 2012 (Bach et.al., 2012) [sic] and this report …[44]

2.2 Objectives, Principles and Procedures, including Implementation and Compliance

2.2.1 Agreement

The Recitals emphasize the cooperative nature of the Agreement in the sustainable development of the Mekong River Basin, which has been extensively studied.[45] Chapter I of the Agreement contains the preamble, preceded by an introductory paragraph indicating that cooperation is to be 'in a constructive and mutually beneficial manner', and to include not just sustainable development, but also sustainable 'utilization, conservation and management'.[46] It is furthermore to be for 'navigational and non-navigational purposes, for social and economic development and the well-being of all riparian States, [and to be] consistent with the needs to protect, preserve, enhance and manage the environmental and aquatic conditions and maintenance of the ecological balance exceptional to this river basin …'.[47] Additionally, the objective is to promote 'independent sub-regional growth and cooperation',[48] 'provide an adequate, efficient and functional joint organisational structure … and to address and resolve issues and problems that may arise … in an amicable, timely and good

---

[43] Above n. 39, 14.

[44] Above n. 39, 2–4.

[45] See also, in addition to references above, Tun Myint, *Governing International Rivers: Polycentric Politics in the Mekong and the Rhine* (Edward Elgar Publishing, Cheltenham, UK and Northampton, MA, USA, 2012); and 'Case Study 1: The Mekong River Basin', in Ariel Dinar, Shlomi Dinar, Stephen McCaffrey and Daene McKinney, *Bridges over Water: Understanding Transboundary Water Conflict, Negotiation and Cooperation* (2nd edn, World Scientific, Singapore, 2013) 273–98.

[46] Introduction to Recitals, Mekong Agreement.

[47] Recitals, Preamble 5. Mekong Agreement.

[48] Recitals, Preamble 6. Mekong Agreement.

neighbourly manner'.[49] Finally, the Agreement is to adhere to the 'objectives and principles of the Charter of the United Nations and international law'.[50]

The 'objectives and principles of cooperation' are the subject of Chapter III of the Agreement. Article 1, entitled 'areas of cooperation', include, but are not limited to 'irrigation, hydro-power, navigation, flood control, fisheries, timber floating, [and] recreation and tourism ...'. Article 1 continues that each of these activities is to be conducted '... in a manner to optimize the multiple-use and mutual benefits of all riparians and to minimize the harmful effects that might result from natural occurrences and man-made activities'. Article 2, entitled 'projects, programs and planning' emphasizes 'joint and/or basin-wide development projects and basin programs through the formulation of a basin development plan'. Article 3, significantly entitled 'Protection of the environment and ecological balance', is directed to 'pollution or other harmful effects resulting from any development plans and uses of water and related resources ...'. Article 4 requires cooperation on the basis of 'sovereign equality and territorial integrity', where 'utilisation and protection of water resources' are at issue. Although not mentioned in any of these articles specifically, Integrated Water Resources Management (IWRM), as considered in the previous chapter in connection with the Aral Sea arrangements, is clearly also highly relevant here. The 2012 report prepared for the MRC highlighted above, explains this process at different levels of application, and emphasizes its significance in the following terms:

> While IWRM at the national level focuses on policies and strategies, water management needs to respect the basin as the basic unit for implementing IWRM on the ground, and hence also for addressing the nexus. To understand the water, energy and food security in a basin (river, lake or groundwater) requires an integrated basin-wide perspective of future development options, together with building knowledge and encouraging dialogue among all stakeholders about the opportunities and risks of more cooperative basin-wide management. This may happen in the context of a formal river basin organisation or through more informal processes. IWRM offers a well-known framework for these discussions in a basin context and is particularly useful because it is implemented at all scales, from watershed though to the transboundary basin.[51]

---

49 Recitals, Preamble 7. Mekong Agreement.

50 Recitals, Preamble 8. Mekong Agreement.

51 Above n. 38, 34. In relation to an example of application to the Mekong Delta in Vietnam, see: Gabi Waibel, Simon Benedikter, Nadine Reis, Sven

Articles 5, 6 and 7 contain more detailed provisions than those preceding them. Article 5 in particular, entitled 'reasonable and equitable utilization' makes reference to the Rules for Water Utilization and Inter-basin Diversion referred to above, and the obligations of Article 5A and B. In relation to Article 5A, intra-basin uses and inter-basin diversions on the tributaries of the Mekong must be notified to the Joint Committee. In relation to the mainstream of the Mekong, different obligations apply depending on whether uses or diversions are within or between basins, and are proposed either in the wet or dry seasons; intra-basin uses – particularly in the wet season – are likely to be less of a concern as flow levels will be higher because of the shared nature of the resource as affected by heavy rainfall.

During the wet season therefore, intra-basin uses must be notified to the Joint Committee, and prior consultations aiming for an agreement by the Joint Committee are required for inter-basin diversions. During the dry season, prior consultation aiming for an agreement by the Joint Committee are therefore also required for intra-basin uses, and for inter-basin diversions, projects must be agreed by the Joint Committee by a specific agreement. The only exception to the requirement for an agreement is if there is a 'surplus quantity of water available in excess of the proposed uses of all parties in any dry season', as 'verified and unanimously confirmed as such by the Joint Committee'. If this is the case, an inter-basin diversion of the surplus can be made subject to prior consultation. The significance of this is considered further below in relation to the Procedures for Notification, Prior Consultation and Agreement.

Article 26, as mentioned above and in the previous section, relates to these obligations. It requires the Joint Committee to prepare and propose for Council approval these rules for the following purposes: to set out the timeframe for the wet and dry seasons; determine where the hydrological stations are to be, and their flow level requirements; outline criteria to determine excess mainstream flow quantities during the dry season; improve monitoring mechanisms for intra-basin use; and to establish a mechanism to monitor mainstream inter-basin diversions. It is clear however that clarification of many of these matters has proved difficult or impossible, in part because of the different views taken by the member states in relation to the hydrological process itself.

Genschick, Loan Nguyen, Pham Cong Huu and Tran Tranh Be, 'Water Governance under Renovation? Concepts and Practices of IWRM in the Mekong Delta, Vietnam', in FG Renaud and C Kuenzer, *The Mekong Delta System: Interdisciplinary Analyses of a River Delta* (Springer, Dordrecht, etc., 2012) 167–198.

In the words of Sneddon and Fox, Thailand, for example, has been inclined to 'overwrite hydrological complexity with geopolitical simplification in order to facilitate autonomy of decision-making'. They go on to state that the representation that underpins Thailand's demands is clearly 'river as watercourse', which makes it possible to 'conceptually separate parts of the river ecosystem'.[52] The problem for Sneddon and Fox is that the principle of equitable utilization, 'the standard method for fostering cooperation in international rivers', is 'far removed from the hydrological and ecological processes of river basins'.[53] Essentially there is a mismatch between international legal obligations and environmental management, with the former failing to understand the ecological dimensions which are essential to the latter.

The maintenance of flows on the Mekong mainstream is the subject of the obligation in Article 6 to cooperate in ensuring that diversions, storage releases, or other permanent actions do not affect this, with the only exception being during times of severe drought and/or floods. Guidelines for locations and levels of flows in relation to this Article must be adopted by the Joint Committee. Article 7 concerns the prevention and cessation of harmful effects, with the due diligence obligation imposed on Parties to 'make every effort to avoid, minimize and mitigate harmful effects that might occur to the environment ...'. The focus is upon water quality, and the condition and balance of the ecosystem. The Article provides that if a riparian state Party is given 'proper and valid' evidence that it is 'causing substantial damage' as a result of its use of, or discharge into the river by another riparian state Party, then such alleged action must 'cease immediately ... until such cause of harm is determined'.

Article 8, 'state responsibility for damages' is the means by which such alleged harmful effects are addressed. In such cases, 'the party(ies) concerned shall determine all relative factors, the cause, extent of damage and responsibility for damages caused by that State in conformity with the principles of international law relating to state responsibility ...'. The means by which this is done is in accord with the dispute resolution provisions in Articles 34 and 35. The first of these provides that the MRC will in the first instance attempt to resolve any dispute in accordance with Articles 18C and 24F. Article 18C simply gives power to the MRC to deal with disputes; Article 24F states that one of the functions of the Joint Committee is to 'resolve issues and differences', and where

52 Sneddon and Fox, above n. 26, 191.

53 Ibid.

necessary 'refer the matter to the Council', so neither provision affects the situation substantively. Article 35 deals with the situation where the MRC is unable to resolve such disputes. In this instance, such matters must be resolved by the governments themselves 'by negotiation through diplomatic channels within a timely manner'. Mediation may also be requested, prior to 'proceed[ing] according to the principles of international law'.

The final two provisions of Chapter III, Articles 9 and 10, concern freedom of navigation and emergency situations respectively. The first ensures freedom of navigation will be protected although it is not prioritized over other uses, so provided dams make provision for the passage of vessels, the right will be upheld. Riparian states remain free to regulate users passing through their own territory in respect of 'sanitary, customs and immigration matters, police and general security' however. The second applies to 'special water quantity or quality problems constituting an emergency', requiring an immediate response. In such cases, where Parties become aware of such situations, they are obligated to consult with other Party(ies) and the Joint Committee so action can be taken.

2.2.2 *Procedures for Notification, Prior Consultation and Agreement*

The *Procedures for Notification, Prior Consultation and Agreement* (PNPCA)[54] provide the means by which activities are carried out in accordance with established practices.[55] The PNPCA essentially provide a framework for EIA in domestic and transboundary contexts.[56] They are explained in related *Guidelines on Implementation* ('the Guidelines'),[57]

---

[54] Mekong River Commission for Sustainable Development, Procedures for Notification, Prior Consultation and Agreement, Approved pursuant to the MRC Council Resolution of 18 October 1999, on the Water Utilization Programme and its subsequent approval by the Joint Committee on 30 November 2003 at its 10th Meeting, Phenom Penh.

[55] See <http://www.mrcmekong.org/> accessed 11/1/16.

[56] For a general overview, see Worawan Sukaroek, 'The MRC Policy Framework on Transboundary EIA: A Better Cooperation will be Enhanced in the Lower Mekong Basin', paper presented to conference entitled *Role of Water Sciences in Transboundary River Basin Management,* Thailand, 2005; and J Horberry, 'Transboundary Environmental Assessment: The Experience of the Mekong River Commission', presentation to the third MOP to the Espoo Convention, 2004.

[57] Mekong River Commission for Sustainable Development, Guidelines on Implementation of the Procedures for Notification, Prior Consultation and

which provide practical assistance as to their content and application.[58] For example, it is stated in relation to EIA, Initial Environmental Evaluations (IEEs), and feasibility studies – and the connections between them – that there are a variety of different approaches in existence:

> … most feasibility study reports already include an EIA or IEE; and in large-scale projects the EIA is prepared in a separate document. It was also noted that the format and content of the feasibility study conducted in one country may be different from another, and feasibility studies carried out for government or local projects may be different from that prepared for international support projects.[59]

The objectives of the PNPCA are to 'provide steps for the MRC member States to support the establishment of the Rules for Water Utilization and Inter-Basin Diversions' (Cl.2a) and to 'promote better understanding and cooperation among the MRC member countries …' (Cl.2b). Notification is the subject of section 4, and Prior Consultation section 5; collectively, these are the main procedures outlined in the document. In connection with the first, intra-basin use and inter-basin diversion on the tributaries, and intra-basin use during the wet season on the mainstream, requires notification to the Joint Committee (Cl.4.1.2), which must be in a timely manner prior to implementation (Cl.4.5).[60] This must include a feasibility study report and an implementation plan, in a form as provided for in Annex I (Cl.4.2). The National Mekong Committee (NMC) of each Party is responsible for handling the notification process, which includes informing relevant agencies, reviewing and checking documentation, and collating and sending the Notification to the Secretariat for submission to the Joint Committee and other NMCs (Cl.4.3). The roles of the Secretariat and Joint Committee are clearly laid out (Cls.4.3.2 and 4.3.3), and if Notification has not been provided, the Joint Committee will request this from the relevant NMC (CL.4.6)

Agreement, approved by the Joint Committee on 31 August 2005 at its 22nd Meeting, Vientiane.

[58] <http://www.mrcmekong.org/assets/Publications/policies/Guidelines-on-implementation-of-the-PNPCA.pdf> accessed 11/1/16.

[59] Ibid, fn 8, p 4.

[60] See also Simon Marsden, 'Developing Agreements for Transboundary Environmental Impact Assessment and Strategic Environmental Assessment in Asia', in Robin Warner and Simon Marsden (eds), *Transboundary Environmental Governance: Inland, Coastal and Marine Perspectives* (Ashgate, Farnham, 2012) 141, 153–156.

Additional data and information under the Prior Consultation process must be provided to allow the other riparian Parties to discuss and evaluate the impact of some proposals. This is needed in three circumstances: for inter-basin diversion from the mainstream during the wet season, intra-basin use on the mainstream during the dry season, and inter-basin diversion of surplus water during the dry season (Cl.5.1). Proposed uses are proposals for a definite use of waters by a riparian, excluding domestic and minor uses not having a significant impact (Cl.1). Prior Consultation attempts to balance the interests of the Parties, and the provisions for this are more detailed than those for Notification, and have different institutional mechanisms. Additionally, 'Specific Agreement' is needed by the Joint Committee in advance for inter-basin diversion projects from the mainstream if during the dry season, with 'terms and conditions such as timing, quantity of diversion, etc.' predetermined (Cl.6).

The NMCs must inform the agencies, review and check information, and collate and send the proposal to the Secretariat, as with Consultations (Cl.5.3.1a–c); however, the MRCs must additionally 'facilitate any consultations, presentations, evaluation and site visit', and record and transmit copies to agencies or Party(ies) 'making the proposal for a definite use of water through the Prior Consultation process of any comments or response received from the MRC Secretariat' (Cl.5.3.1d–e). The Secretariat and Council also have additional roles to play regarding Prior Consultation. After the Secretariat has reviewed, analysed and provided technical advice, supplied additional data and facilitated any meetings requested, provided technical support for any evaluation and 'supported any fact-finding team', and entered data on the MRC Data Information System (Cl.5.3.2c(f)), the Joint Committee essentially evaluates any potential significant environmental effects, and can establish a Working Group following an evaluation of the information from the notifying state by each of the other notified states (Cl.5.3.3c–e).

This process for Prior Consultation is elaborated more fully in Cl.5.4. Three stages are involved: submission of documents, evaluation and reply to proposed use, and the decision by the Joint Committee. The Guidelines provide more information on these stages in Prior Consultation as well as for Notifications. In relation to evaluations for example, the Joint Committee can direct the Secretariat, or appoint a working group or technical advisory team to assist (Cl.5.4.2). Decisions by the Joint Committee on the proposed use should contain any agreed conditions (Cl.5.4.3). Additionally, the notifying State must 'not implement the proposed use without providing the opportunity for the other member

States to discuss and evaluate the proposed use' (Cl.5.4.3). Sneddon and Fox explain the significance of the different consultation requirements as follows:

> Prior consultation is the stricter of the two standards because it stipulates that an actual dialogue among the riparian states must occur before a water project can proceed. Presumably, a particularly objectionable project (for example, one that results in dry season, main channel flows below the minimum flow threshold) would meet its demise during the consultation process. On the other hand, notification, or simply informing one's riparian neighbours of one's intentions regarding a water resource development project, is the less demanding standard and is required for intra-basin uses during the wet season, and for alterations such as impoundments and diversions on the tributaries.[61]

One of the weaknesses of the Procedures is the absence of any obligations for public participation.[62] The MRC does have a public participation strategy, but in practice it is distant from the public.[63] This needs to change so that the NMCs 'can engage with civil society and the Basin public within each country'.[64] Writing in 2003, Dore explains the complex reasons that there had been little achievement in this area as: 'The youth of the new version of the organisation, the sustainability orientation and mindset of some of the agencies which dominate the National Mekong Committees, limited successes to this point and operating rules which limit engagement with the wider basin community …'.[65] A legal basis for the

---

[61] Above n. 26, 190–191.

[62] Jona Razzaque, 'Public Participation in Water Governance', in Joseph W Dellapenna and Joyetta Gupta (eds), *The Evolution of the Law and Politics of Water* (Springer, 2009) 353, 359.

[63] A study on public participation in the context of the MRC was initiated in 1996, a report completed – with guidelines – in 1998, and approved by the Joint Committee in 1999. Despite this, it does not appear to have met the concerns of the critics, such as Dore, above n. 30, 17. See: <http://www.mrcmekong.org/assets/Publications/policies/Public-Participation-in-MRC-context.pdf> accessed 1/4/16.

[64] See Philip Hirsch and Kurt Mørck Jensen with Ben Boer, Naomi Carrard, Stephen FitzGerald and Rosemary Lyster, *National Interests and Transboundary Water Governance in the Mekong* (Australian Mekong Resource Centre, University of Sydney, Danish International Development Assistance, 2006), in particular Chapter 3: 'Legal and Institutional Framework for Mekong Water Governance' 26, 122.

[65] Above n. 30, 17.

transboundary EIA process in relation to proposed inter-basin and intra-basin diversions (number 7), and mandatory requirements for public participation (number 8), are recommended by Hirsch et al. who comment:

> A further issue in the context of formulating binding procedures and rules under the Agreement concerns transboundary environmental impact assessment. While basin-wide environmental and social impact assessment processes have been in place since 1998, there appear to be no legally-based standards for conducting those assessments. It would be desirable to formulate such standards and include them in rules or in a protocol to the Agreement. An important starting point for drafting such mechanisms would be the *Convention on Environmental Impact Assessment in a Transboundary Context*, which obliges signatories to 'take all appropriate and effective measures to prevent, reduce and control significant adverse transboundary environmental impact from proposed activities' and to 'take legal and administrative measures to implement these provisions.'[66]

However, national implementation and compliance generally remains weak and ineffective as vigorously noted by Contreras:

> While the signatory countries in the Mekong subregion participated in the drafting of several procedures embedded in the agreement, there is no evidence that they have made any reference to those procedures or other provisions of the agreement in their national water policies and laws, including those they have passed after the agreement came into force. This stems not only from the lack of strong compliance provisions in the Mekong regional agreement, but also from the weak, if not incoherent, water legislation in each of the signatory countries.[67]

In relation to evidence of practice, it can be noted that funded by the Global Environment Facility (GEF)[68] and executed by the WB, in 1999 the MRC and member states adopted a Water Utilisation Program to develop rules for water utilization. The weakness of the MRC is however, in the view of one author '… compromised by structural limitations. Like ASEAN, the MRC has no means by which to push member countries to harmonize their national policies according to the agreements forged'.[69] Most of the activities impacting the Mekong River are of course operations connected with dam building, the majority of which have been conducted outside the framework of the agreement, with little or no

---

[66] Above n. 64, 31. Rieu-Clarke below n. 74 also makes reference to the Espoo Convention, 157–158, 164–165; and see Marsden, above n. 60, 153.

[67] Contreras, above n. 7, 164.

[68] <https://www.thegef.org/gef/> accessed 18/7/16.

[69] Contreras, above n. 7, 164.

transboundary EIA, and inadequate public participation of those affected by the development. Early examples include the Vietnamese Sesan Dam, the Thai water grid, the Laos Thuen Dam and Chinese development on the Upper Mekong.

A Danish-sponsored study criticized the flexibility of the rules, lack of enforcement, adherence to the participation provisions, and motivation and/or political will by the MRC and its Parties, with national interests dominating outcomes.[70] The struggle over the Thai Pak Mun Dam is also cited with some distain as 'exemplary of those water conflicts that occur within the national territories of the Mekong riparian states, but remain generally invisible to the institutional arrangements constituted by the Mekong Agreement'.[71] This saw fishing households displaced, with loss of their traditional livelihoods, largely to develop the region at the behest of the US to avoid the spread of cross-border communist influence from Cambodia.

The reasons why the obligations of the *Mekong Agreement* had no real impact upon the decision to proceed with the Pak Mun Dam appear to be closely tied with the future aspirations of Thailand within southeast Asia; in particular its lack of willingness to cede sovereignty over what was regarded as a 'national' project from the start, which was intended to develop the northeast of the country with little if any regard for its transboundary effects. Sneddon and Fox highlight the geopolitics of southeast Asia, whereby Cambodia's withdrawal from the then Mekong Committee in 1975 assisted with the nation-building focus of the project. This ensured there were no obligations under the Agreement to neighbouring Cambodia, and the effects upon the subregion as a whole were thereby largely hidden.[72] The involvement of the WB in financing energy development in Thailand helped further in limiting consideration of the

---

70 For further analysis, see Fleur Johns, Ben Saul, Philip Hirsch, Tim Stephens and Ben Boer, 'Law and the Mekong River Basin: A Socio-Legal Research Agenda on the Role of Hard and Soft Law in Regulating Transboundary Water Resources' (2010) 11(1) *Melbourne Journal of International Law*, 154–174; and Ben Boer, Philip Hirsch, Fleur Johns, Ben Saul and Natalia Scurrah, *The Mekong: A Socio-legal Approach to River Basin Development* (Routledge, Abingdon, 2015).

71 Sneddon and Fox, above n. 26, 193.

72 Sneddon and Fox, above n. 26, 194. See in relation to Cambodia, Oliver Hensengerth, 'Global Norms in Domestic Politics: Environmental Norm Contestation in Cambodia's Hydropower Sector' (2015) 28(4) *The Pacific Review* 505–528

project as a national rather than a subregional one, which was aided by the Executive Directors (EDs) of the Bank, as Sneddon and Fox comment:

> The Pak Mun project was structurally tied to the larger program of energy development that the Thai state sought to fund via the World Bank. The scalar imagination of the World Bank, as revealed through the EDs' comments, was perforce limited to the scale of the nation as a whole, a view that meshed perfectly with the perspective of the Thai state. Accordingly, the Pak Mun conflict is largely invisible and unheard from the vantage point of the transnational mechanisms to govern the Mekong ...[73]

While the *Mekong Agreement* may have had little application in the Pak Mun project, the consultation procedures under the Agreement were applied in the Xayaburi hydropower project, but despite this found to be largely ineffective. In 1994 the MRC had identified the Xayaburi project as one of a dozen potential dams on the mainstream of the Lower Mekong. As the first to be subject to the PNPCA, it has been examined extensively, with Rieu-Clarke finding that 'the Xayaburi prior consultation process failed in its main objective, namely to reconcile the competing interests of the states concerned'.[74] He finds that 'this failure to reconcile competing interests can partly be blamed on the ambiguity within the text of the 1995 Mekong Agreement and related procedures'.[75] He furthermore specifies the 'key areas of ambiguity' as 'when the notification should take place, what data and information should be shared, the requisite time period for consultations, whether the planned project could proceed during the consultation period, and what should be the next steps if the states failed to agree at the end of the consultation period'.[76]

---

73 Sneddon and Fox, above n. 26, 196.

74 Alistair Rieu-Clarke, 'Notification and Consultation Procedures under the Mekong Agreement: Insights from the Xayaburi Controversy' (2015) 5 *Asian Journal of International Law* 143, 145. See also Oliver Hensengerth, 'Where Is the Power? Transnational Networks, Authority and the Dispute over the Xayaburi Dam on the Lower Mekong Mainstream' (2015) 40(5–6) *Water International* 911–928.

75 Ibid.

76 Above n. 74, 146, and specifically: 156–159 'timely notification' (with reference to the Espoo Convention and general international law); 159–165 'available' and 'relevant' data and information; 165–168 the prior consultation period; 168–173 the suspension of the project during the consultation period; and 173–174 the post-consultation procedures.

In April 2011, at a special session of the Joint Committee, the MRC deferred a decision on the Xayaburi hydropower plant in Laos. Concerns had been raised by Cambodia, Thailand and Vietnam over gaps in technical knowledge, the predicted environmental impact, effect upon the livelihoods of the inhabitants of the Basin, and the need for broader consultation. Vietnam furthermore recommended deferring all development for at least a decade while environmental concerns continued to be expressed. The decision was informed by the MRC Final Report on *SEA of Hydropower on the Mekong Mainstream*.[77] This was a proactive effort to evaluate all 12 hydropower schemes, eight of which were in Laos.[78] Despite the deferral decision and SEA Report, construction work on the Xayaburi Dam was commenced by Laos in the absence of a decision by the MRC or adequate consultation. Notwithstanding Thailand's concerns, as it is to be the consumer of the electricity produced, financing of the project was perhaps not surprisingly provided by six Thai banks.

Most recently, consultation on another major project, the Don Sahong Hydropower Project – again in Laos – commenced in July 2014. Given the inadequate consultation over the Xayaburi Dam, in September 2014 an environmental NGO – Save the Mekong Coalition – sent a letter to the prime ministers of the MRC Parties, urging them to halt the consultation over the Don Sahong project and reform the PNPCA process first.[79] This was because other than the MRC Parties, there was no formal means under the PNPCA procedures for gathering and inputting the concerns of other (non-government) stakeholders. The MRC agreed to hold a meeting in relation to this in December 2014, and based on the outcomes, the Joint Committee agreed to report on the need to reform the PNPCA process.[80]

Promoting sustainable development of the Mekong Basin ultimately requires choices to be made to take account of the economic, socio-political and ecological aspirations of the member countries to the *Mekong Agreement* and the communities that depend on the Basin for their livelihoods. Without linking water rights with social justice, sustainable development is unlikely, and despite its frequent ineffectiveness as a

---

77 Above n. 37.

78 See Marsden, above n. 60, 156.

79 Save the Mekong Coalition, 'Request for Clarification Regarding the First Regional Consultation on the Don Sahong Hydropower Project', Letter to Mr Hans Guttman CEO of the Mekong River Commission, 11 November 2014.

80 <http://www.mrcmekong.org/news-and-events/news/mrc-holds-special-session-on-prior-consultation-of-don-sahong-hydropower-project/> accessed 11/1/16.

regime, the *Mekong Agreement* has an important role in this.[81] Ponte highlights this and the need to address the 'three key determinants' to meet these aspirations: 'the rural inhabitants of the Mekong River Basin, the ecosystem's health, and the river flow regime'. She also highlights climate change as presenting a 'major challenge in the Mekong region'.[82] While the use of hydro-electricity may be a significant driver to alleviate poverty and improve development outcomes, there are significant downsides that, if not managed carefully and effectively, may alter river flow to the major detriment of community livelihoods.

3. THE CONSERVATION AND CENTRE FOR BIODIVERSITY AGREEMENTS

3.1 Geographical Context, Historical Background and Institutional Arrangements

Southeast Asia's location in the tropics, together with its large and growing population and associated development, has placed major pressures on the natural environment, and particularly the land hosting the vast biodiversity of plants and animals residing there. In addition to being home to around half a billion people, the area is one of the most significant places globally for nature conservation, supporting a sixth of the world's mammals and amphibians, and almost a quarter of the world's birds. The IUCN also plays an important part in this subregion.[83] An IUCN Southeast Asia Group was established in 2011 to support the growth of its work in the subregion, '... particularly through multi-country/trans-boundary interventions, and to provide a platform for IUCN interaction with Regional bodies such as ASEAN'.[84]

On 9 July 1985, at the Eighteenth Annual Meeting of ASEAN, the Foreign Ministers of Brunei, Indonesia, Malaysia, the Philippines, Singapore and Thailand adopted the *Conservation Agreement*. The institutional arrangements for the *Conservation Agreement, Centre for Biodiversity*

---

[81] See Kate Lazarus, Nathan Badenoch, Dao Nga and Bernadette P Resurreccion (eds), *Water Rights and Social Justice in the Mekong Region* (Earthscan, London, 2011), where perhaps for this reason there is virtually no analysis of the Mekong Agreement.

[82] Above n. 22, 19.

[83] <https://www.iucn.org/commissions/commission-ecosystem-management/regions/south-east-asia> accessed 14/11/16.

[84] Ibid.

Agreement (and the *Haze Pollution Agreement*) are overseen by ASEAN, the primary body responsible for the environmental governance of the subregion, which was outlined above. However, since the adoption of the *Conservation Agreement*, another specific body has been created, which 'embodies the commitment of ASEAN Member States in establishing ACB [the ASEAN Centre for Biodiversity] as a regional centre that facilitates cooperation and coordination among ASEAN Member States and with relevant organizations on the conservation and sustainable use of Southeast Asia's rich but highly threatened biodiversity'.[85] Established under a related treaty, the *Centre for Biodiversity Agreement*, the purpose of the European-Commission-sponsored ASEAN Centre for Biodiversity,[86] is to 'facilitate cooperation and coordination among the members of ASEAN' (Article 2). Curiously, but perhaps because it is not in force, without reference to the *Conservation Agreement* which preceded it by two decades, the *Centre for Biodiversity Agreement* describes itself as: 'as the first regional initiative to save the ASEAN region's critically threatened biodiversity'.[87] The institutional arrangements for the Centre are outlined in section 3.2 below.

3.2 Objectives, Principles and Procedures, including Implementation and Compliance

3.2.1 *Conservation Agreement*

The *Conservation Agreement* remains a non-binding instrument containing many broad environmental objectives and obligations.[88] There are eight chapters to the Agreement, made up of 35 Articles, with the Chapters entitled as follows: Conservation and Development (Chapter I); Conservation of Species and Ecosystems (Chapter II); Conservation of Ecological Processes (Chapter III); Environmental Planning Measures (Chapter IV); National Supporting Measures (Chapter V); International Cooperation (Chapter VI); International Supporting Measures (Chapter

---

[85] <http://www.aseanbiodiversity.org/index.php?option=com_content&view=article&id=489:thailand-ratifies-acb-establishment-agreement&catid=1:news&Itemid=109> accessed 18/1/16.

[86] Ibid: the EC approved in 2004 a grant of €6.0 million to support the establishment and initial operation of a new institution.

[87] Ibid.

[88] See Koh Kheng-Lian, 'ASEAN Environmental Protection in Natural Resources and Sustainable Development: Convergence versus Divergence?' (2007) 4(1) *Macquarie Journal of International and Comparative Environmental Law* 43–73.

VII); and Final Clauses (Chapter VIII). The fundamental principle (Article 1(1)) is that Parties agree to adopt the necessary measures 'to maintain essential ecological processes and life-support systems, to preserve genetic diversity, and to ensure the sustainable utilisation of harvested natural resources'. In accordance with Article 1(2), national conservation strategies are to be developed in support of these measures and coordinated within a regional conservation policy framework. Despite not being in force, Elliott comments on its significance as follows:

> While, at the time of writing, the Agreement is not in force (because the required six ratifications have not yet been received) it nevertheless signalled a transition into a new phase of cooperation in which environmental factors were to be given equal consideration with economic and social ones. Among other things, it recognized that member states had a legal obligation to minimize cross-boundary environmental impacts. It set out principles by which shared resources were to be managed and transboundary environmental effects such as pollution avoided.[89]

The *Conservation Agreement* includes specific provisions on using environmental planning measures (through land use planning and protected areas), giving 'as full consideration to ecological factors as to economic and social ones' (Article 2(2)); conserving genetic diversity, 'by taking action aimed at ensuring the survival and promoting the conservation of all species' (Article 3(1)), to include conservation of habitats, sustainable use of harvested species, and endangered and endemic species, and 'taking all measures' to prevent the extinction of species (Article 3(2)). The creation and maintenance of protected areas, the regulation of the taking of species and introduction of exotic species, and the promotion and establishment of collections of genetic resources are a key part of these obligations (Article 3(3)).

The sustainable use of harvested species is the subject of Article 4, which requires the development, adoption and implementation of management plans to prevent any decrease in the size of any harvested population, maintain ecological relationships, restore depleted populations, and avoid changes in the ecosystem which are not reversible. However, the regulatory approach is limited by 'appropriate and necessary legislative and administrative measures on harvesting activities' being subject to the 'national interests' of the member states (Article 4). Endangered and endemic species is the subject of Article 5, with the

---

[89] Elliott, above n. 15, 61–62.

former listed in Appendix I. The taking of these species is prohibited, trade in and possession of them is regulated, and there is an obligation to protect their habitat 'wherever possible' (Article 5(1)). In relation to species endemic in areas under the jurisdiction of any state Party, 'all the necessary measures to maintain the population of such species at the highest possible level' is required (Article 5(3)).

Vegetation cover and forest resources is the subject of Article 6 of the Agreement; soil, Article 7; water, Article 8; and air, Article 9. The focus of the conservation of forest cover is upon the control of vegetation clearance, the prevention of bush and forest fires, and the prevention of overgrazing (Article 6(2)(a)). It is also about the regulation of mining to prevent vegetation clearance, setting aside areas for reserves, avoiding monocultures and maximizing species diversity, protecting soils and developing management plans based on ecological principles (Article 6(2)(b–g). The focus on soils in Article 7 is designed to 'to prevent soil erosion and other forms of degradation, and promote measures which safeguard the processes of organic decomposition'. Measures to be taken for this purpose include the development of land use policies (Article 7(a)), taking all measures to control erosion (Article 7(b)) – linked with vegetation cover), and taking 'appropriate measures to rehabilitate eroded or degraded soils' (Article 7(c)).

The conservation of underground and surface water resources is the focus of Article 8, with Parties required to undertake research into watershed characteristics (Article 8(2)(a)) and to regulate and control water utilization (Article 8(2)(b)), in addition to an obligation of EIA, 'when planning and carrying out water resource development projects [to] take fully into account possible effects of the projects on natural processes or on other renewable natural resources and prevent or minimize such effects' (Article 8(2)(c)). Finally, in relation to air – which is symptomatic of the position that many southeast Asian states find themselves in thirty years after the adoption of this treaty – Parties are required to 'take all appropriate measures towards air quality management *compatible with sustainable development*' (Article 9, my emphasis). The failure to ratify the *Conservation Agreement* and implement its provisions clearly led to the need for the *Haze Pollution Agreement*, which is analysed below.

The *Conservation Agreement* also has provisions on conserving ecological processes which are found in Articles 10 (Environmental Degradation) and 11 (Pollution). In relation to the former, Parties are obliged to promote environmentally sound agricultural practices, pollution control and the development of environmentally sound industrial processes, and adequate economic or fiscal incentives to achieve the foregoing (Article

10(a–c)). The Article is based on the polluter pays principle – and associated preventative and user pays principles – that the person responsible must prevent, reduce and control the activity, and where possible, rehabilitate and remediate (Article 10(d)). Parties must also consider the cumulative effects of authorizing actions upon those already taking place (Article 10(e)); and 'pay particular attention' to 'adverse effects on processes which are ecologically essential or on areas which are particularly important or sensitive' (Article 10(f)). The control of pollution is the focus of Article 11(a), with cumulative effects and the 'self-purificating aptitude of the recipient natural environment' – the receiving environment – to be taken into account. Treatment of pollution is to be a condition of any regulation (Article 11(b)), and national environmental quality monitoring programmes must be established (Article 11(c)).

Where possible, natural resource conservation must be integrated into land use planning processes in accordance with Article 12. In the interests of sustainable development, this means that Parties must 'give as full consideration as possible to ecological factors as to economic and social ones' (Article 12(1)), and this provision also requires that land use plans must be based on ecological carrying capacity; in particular, 'the importance of retaining the naturally high productivity of areas such as coastal zones and wetlands' must be considered (Article 12(2)). Where natural resources are of importance to more than one Party – such as transboundary protected areas – efforts must be coordinated (Article 12(3)).

Article 13 concerns protected areas specifically, and requires the establishment of 'terrestrial, freshwater, coastal or marine protected areas' by Parties in their own jurisdictions; these must safeguard ecological and biological processes, representative samples, satisfactory population levels, and areas of particular importance (Article 13(1)). The Article requires furthermore that Parties must:

> … take all measures possible in their power to preserve those areas which are of an exceptional character and are peculiar to their country or the Region as well as those which constitute the critical habitats of endangered or rare species, of species that are endemic to a small area and of species that migrate between countries of Contracting Parties.

The regional perspective is very important with respect to the creation of protected areas, with Article 13(6) requiring that:

> Parties shall co-operate in the development of principles, objectives, criteria and guidelines for the selection, establishment and management of protected

> areas in the Region with a view to establishing a coordinated network of protected areas throughout the Region, giving particular attention to those of regional importance.

Protected areas must be designated either as national parks or reserves, the former 'to allow for ecological self-regulation of one or several ecosystems' and the latter to be 'set aside for the purpose of preserving a specific ecosystem' (Article 13(3)); each is to be 'regulated and managed in such a way as to further the objectives for the purpose of which they have been created' (Article 13(2)). Protected areas are not limited to public lands however, and further to Article 13(7), 'Parties shall promote, through the adoption of appropriate measures the conservation of natural areas by private owners, community or local authorities'. Management plans are to be prepared for all protected areas, including buffer zones prohibiting or regulating uses (Article 13(4)). Once protected areas are established, the introduction of exotic species, and use or release of toxic substances or pollutants must be prohibited, and uses outside the areas that may impact upon them detrimentally prohibited or controlled (Article 13(5)).

Article 14 contains the (brief) obligation to conduct EIA 'for any activity which may significantly affect the natural environment' prior to its adoption, and to 'take into consideration the results of this assessment in their decision-making process' (Article 14(1)). Where activities are carried out, they must be done so 'to overcome or minimize any assessed adverse effects and shall monitor such effects with a view to taking remedial action' (Article 14(2)). The introduction of national support measures (for example scientific research and public participation), is the subject of Articles 15 and 16. Article 17 requires Parties to establish and fund designated authorities to carry out obligations under the Agreement.

The obligation is imposed on Parties to coordinate their activities to achieve the purpose of the Agreement under Article 18. This involves a number of things, including: monitoring, research, exchange of information, and consultation and assistance regarding implementation (Article 18(2)). Article 18(3) furthermore requires Parties to forward to the Secretariat information on the 'biological status of the natural living resources of the Region'. Under Article 19, Parties with shared natural resources are to 'cooperate concerning their conservation and harmonious utilisation', so as to avoid adverse environmental effects upon each other. EIA, advance notification and appropriate consultation should all be undertaken prior to any proposed utilization of the shared resources by one of the Parties (Article 19(2)). Parties must also 'endeavour to conclude bilateral or multilateral agreements in order to secure specific regulation of their conduct in respect of the resources concerned' (Article 19(2)(b)).

The obligation therefore sets out requirements of one state in relation to another (Article 19(2)(c)–(g)), and Article 20 reinforces this by referring specifically to 'transfrontier environmental effects', detailing requirements for transboundary EIA (Article 20(3)), reminding Parties of the international legal principle of state responsibility, and calling on them to avoid 'to the maximum extent possible' causing adverse environmental effects beyond their national boundaries (Article 20(2)). Furthermore, 'Parties shall, in particular, endeavour to refrain from actions which might directly or indirectly adversely affect wildlife habitats situated beyond the limits of national jurisdiction, especially habitats of species listed in Appendix I or habitats included in protected areas' (Article 20(4)).

The *Conservation Agreement* is clearly a comprehensive arrangement to address environmental governance in southeast Asia, and while this is perhaps its greatest strength, it appears also to have been its greatest weakness. The reason it is not in force can therefore be linked directly with this fact, as well as other developments in environmental law and policy in the 1990s, which in one sense may have rendered it obsolete. The diverse nature of the ASEAN states and different levels of development have also led to some questioning the viability of the Agreement. Writing a decade ago (2007) the leading scholar of ASEAN environmental law and policy, Koh Kheng-Lian, noted that a call was made to assess the relevance of the Agreement in the light of a number of MEAs, but nothing appears to have happened since. She summarized the challenges posed by the Agreement in the following terms, raising the potential impediment of the 'ASEAN Way' at the end:

> The scope is very wide and was perhaps too forward-looking at that period of time for member states to implement, as it specified an inventory of flora and fauna among other requirements. Subsequently, when the 1992 United Nations *Convention on Biological Diversity* came into existence, the question was posed as to whether it had overtaken the Agreement in terms of its contents. Was there still a need for the Agreement? Have there been other ASEAN initiatives that have now rendered the Agreement outdated, otiose or replaced many of its provisions? Should each ASEAN member state proceed with its ratification of global conventions and implement those in accordance with the 'common but differentiated responsibilities' principle? Or, should ASEAN work towards a harmonized biodiversity strategy under this Agreement, together with other ASEAN strategies, programs and plans of action?

> Has the 'ASEAN Way' to cooperation among its members been an obstacle to the hard law approach at the subregional level?[90]

The final provisions of the *Conservation Agreement* relate primarily to administrative arrangements together with dispute settlement. Article 21 sets out the provisions for the MOP, to be held at least once every three years, in conjunction with ASEAN meetings where possible (Article 21(1)). The function of the MOP is to review implementation; adopt, review and amend appendices; consider reports from Parties; make recommendations concerning Protocols or amendments of the Agreement; and establish working groups or other subsidiary bodies (Article 21(2)). Similarly, Article 22 concerns the administrative role of the Secretariat, and Article 23 makes provision for National Focal Points to channel communications to the Secretariat from the Parties. The final clauses enable adoption of Protocols (Article 24), amendment (Article 25), and provisions for appendices and amendment thereto (Article 26). The dispute settlement provision is remarkably brief, and consistent with the 'ASEAN Way', stating: 'Any dispute between the Contracting Parties arising out of the interpretation of implementation of this Agreement shall be settled amicably by consultation [o]r negotiation' (Article 30). The benefits of the *Conservation Agreement* have recently been summarized as follows:

> The main strength of the 1985 ASEAN Conservation Agreement is in the content of its provisions, because it addresses important concepts and principles that serve as prerequisites in achieving sustainable development. Instead of simply individually protecting individual animals and plants … the 1985 ASEAN Conservation Agreement aims to preserve the entire ecosystem and emphasizes the importance of genetic diversity. Its carefully crafted preamble integrates all the ingredients of a good environmental agreement such as inter-generational equity, balance between conservation and socio-economic development, and the importance of economic cooperation.[91]

3.2.2 *Centre for Biodiversity Agreement*

The *Centre for Biodiversity Agreement* was adopted to facilitate biodiversity cooperation across ASEAN and perhaps is an indication of the changed focus since the effective stalling of the *Conservation Agreement.*

---

[90] Above n. 88, 53.

[91] S Jayakumar, Tommy Koh, Robert Beckman and Hao Duy Phan (eds), *Transboundary Pollution: Evolving Issues of International Law and Policy* (Edward Elgar Publishing, Cheltenham, UK and Northampton, MA, USA, 2015) 311.

The Centre itself became an international organization after Myanmar ratified the Establishment Agreement on 8 July 2009. Following Brunei, Laos, the Philippines, Singapore and Vietnam, Myanmar was the sixth ASEAN Member State to ratify the Agreement. The Philippines subsequently became the headquarters of the Centre with the ratification of the Host Country Agreement signed between the Centre and the Government of the Philippines on 8 August.[92] The significance of the international legal personality of the Centre is described as follows: 'Becoming a legitimate international organization will enable the Centre to mobilize more resources and forge more partnerships that will result in enhanced support to the ASEAN Member States in meeting their commitments to various multilateral environmental agreements and in reducing the rate of biodiversity loss'.[93]

The establishment of the Centre is confirmed by Article 1 of the Agreement, with the purpose acknowledged to be 'to facilitate cooperation and coordination among the members of ASEAN, and with relevant national governments, regional and international organisations, on the conservation of biological diversity and the fair and equitable sharing of benefits arising from the use of such biodiversity in the ASEAN region' (Article 2). The structure comprises a Governing Board, Executive Director, and committees and subsidiary bodies (Article 3). The Governing Board consists of member of ASEAN Senior Officials on the Environment and the Secretary General of ASEAN (Article 4(2)); observers may be invited as appropriate (Article 4(3)). The Board is to meet at least once a year (Article 3(5)), and must report to the ASEAN environment ministers (Article 3(6)).

A broad range of administrative functions is given to the Board, including the provision of policy guidance and directions, and overall supervision of the Centre (Article 5(a)); approving rules, procedures and the business plan, and administering funds are all provided for in related sub-sections (Article 5(b)–(h)). An Executive Director appointed by the Board is to be responsible for day-to-day operations (Article 6); in turn a range of duties are specified for the Director to perform, including administering the Centre and its programmes, preparing annual work programmes and plans, and reporting on implementation (Article 6(5)). The ASEAN Biodiversity Fund is established under Article 8, with voluntary contributions from member states (Article 8). Legal personality

---

92 Host Country Agreement between the Government of the Republic of the Philippines and the ASEAN Centre for Biodiversity (Manila, 8 August, 2006).

93 <http://www.aseanbiodiversity.org/index.php?option=com_content&view=article&id=69&Itemid=79&current=1> accessed 19/1/16.

is granted by Article 9, and any disputes concerning interpretation or implementation are to be settled amicably through consultation or negotiation in accordance with Article 12.

4. THE HAZE POLLUTION AGREEMENT

4.1 Geographical Context, Historical Background and Institutional Arrangements

Haze pollution is explained by particulate matter in the air resulting from smoke from fires, often combined with dust. It occurs in southeast Asia typically from May to September when humidity is at its lowest. Ground-level visibility is often reduced to between 1000 and 2000 metres, and in the absence of wind conditions can remain the same for weeks or even months at a time. As with other forms of air pollution, human health and livelihoods are detrimentally affected and the large number of people affected multiplies the problem significantly. The origin of the fires is mostly Indonesia, primarily the island of Sumatra, where it is caused by burning trees and draining peatland to create agricultural land. Some agriculture is based around small-scale activity, but industrial-scale palm oil plantations (mostly for export to India and China) and logging of tropical timber are the major motivating causes.[94] While some fires may begin by lightning strikes, the majority are, however, deliberately lit during clearance activity, which has been described as follows:

> Slash-and-burn agriculture, which relies on fire to clear otherwise seemingly indestructible vegetation, is a traditional practice for clearing land in Southeast Asia, especially in Indonesia. Plantation owners and farmers alike prefer the slash-and-burn technique because it is cheap, easy and effective – just light a match and control the burn. The indiscriminate use of this method,

---

[94] For further detailed background, see: H Litta, 'Case Study One: Transboundary Haze Pollution', in *Regimes in Southeast Asia: An Analysis of Environmental Cooperation* (Verlag Springer, Wiesbaden, 2012) 89-102-103 (in relation to palm oil development). See also Helena Varkkey, *The Haze Problem in Southeast Asia: Palm Oil and Patronage* (Routledge, Abingdon, 2016), who addresses the palm oil issue specifically.

however, destroys the vegetation that covers peatlands, leaving them exposed and vulnerable to fire.[95]

In 2002, in an attempt to deal with the issue, ASEAN negotiated the *Haze Pollution Agreement*, which entered into force the following year after the required six states had ratified.[96] The historical origins can be traced to incidents of 'problematic haze levels' occurring in 1982 and 1987,[97] followed by a turning point in the regional haze crisis in 1997, which caused unprecedented health and other effects in southeast Asia. Goodsite et al. note that over half a million premature deaths a year in Asia may be caused by air pollution, with millions more suffering from respiratory illnesses, with 2–4 per cent of the region's GDP wiped out as a result of loss of productivity and healthcare costs alone; haze pollution is a major contribution to this.[98] The fires on Sumatra (and Kalimantan) in 1997 were exacerbated by a severe drought, and the wind direction meant the haze affected Indonesia, Malaysia, Singapore, Brunei, Thailand, and even the Philippines.[99] Another particularly bad year was 2015, when the pollution reached the southern Philippines island of Mindanao, over 1200kms away, and lasted until as late as the end of October.[100] Jerger describes the international monitoring activity undertaken by the USA in 1997:

> Throughout the crisis, the U.S. National Oceanic and Atmospheric Administration monitored the affected areas. Schools and offices shut down, planes crashed, and people died from acute respiratory failure. By the time the fires

[95] David B Jerger, 'Indonesia's Role in Realizing the Goals of ASEAN's Agreement on Transboundary Haze Pollution' (2014) 14(1) *Sustainable Development Law and Policy* 35, 36.

[96] See generally Paruedee Nguitragool, *Environmental Cooperation in Southeast Asia: ASEAN's Regime for Transboundary Haze Pollution* (Routledge, London, 2011), and as an example of one of several comparisons, this with the EU arrangements, see: Michael Goodsite, Catherine Mei Ling Wong, Ole Hertel and Lars Moseholm, 'Responses to Air Pollution Based on Historical and Current Policies in the EU and ASEAN' (2011) 6 *Global Environment* 150, 165–173.

[97] Goodsite et al., ibid, 166.

[98] Ibid, 167.

[99] The relationship between the countries causing the pollution and those affected by it is explored by Litta, above n. 94, at 103–104.

[100] ABC online, 'Haze from Indonesian Forest Fires Reaches Southern Philippines, Causes Air Travel Disruptions' 24 October 2015. See: <http://www.abc.net.au/news/2015-10-24/indonesian-haze-reaches-the-philippines/6881564> accessed 8/4/16.

> were brought under control, the region had suffered widespread forest destruction, losing nearly ten million hectares. The fires destroyed portions of seventeen protected forest areas in Indonesia and land that could have otherwise been used for agriculture.[101]

Nine countries have now ratified the Agreement: Brunei, Cambodia, Laos, Malaysia, Myanmar, the Philippines, Singapore, Thailand and Vietnam. According to ASEAN, it is the 'first regional arrangement in the world that binds a group of contiguous states to tackle transboundary haze pollution resulting from land and forest fires'.[102] Notwithstanding the implications of the 'ASEAN Way' for implementation and compliance, the absence of Indonesian ratification remains however the principal impediment to its success, as noted by Mushkat who comments: 'Indonesia has been the most glaring absentee in that, without its active participation, a new governance regime could not function adequately'.[103] However, Mushkat indicates that the structural issues within Indonesian governance, what she terms the 'pervasive fragmentation of the environmental management system',[104] mean that even if Indonesia were to ratify the Agreement, this would hold no guarantee of success:

> The problem has two principal dimensions. At the central government level, multiple bureaucratic entities perform vaguely delineated and overlapping functions with virtually no coordination and synchronized planning. In addition, far-reaching administrative decentralization has rendered them effectively impotent vis-à-vis the plethora of largely autonomous sub-national units engaged in activities which impinge on ecological governance. The corollary is that the country may not be in a position to cooperate meaningfully with its neighbours in the environmental domain, whatever its international commitments.[105]

The *Haze Pollution Agreement* builds on, or gains inspiration from, a number of arrangements. The first is the Regional Haze Action Plan (RHAP), which was said to have 'furnished a conceptual framework for

---

101 Above n. 95, 40, footnotes deleted.

102 ASEAN, 'Haze Action Online – ASEAN Agreement on Transboundary Haze Pollution', available 13 November 2014 at <http://haze.asean.org/?page_id=185> accessed 11/1/16.

103 Roda Mushkat, 'Creating Regional Environmental Governance Regimes: Implications of Southeast Asian Responses to Transboundary Haze Pollution' (2013) 4(1) *Washington and Lee Journal of Energy, Climate and the Environment* 103, 145.

104 Ibid.

105 Ibid, 145–146.

combating transboundary pollution';[106] the second is the approach of the UNECE *Air Pollution Convention*, noted by Jerger;[107] and the third is the ASEAN *Conservation Agreement*, which, while not focused specifically on transboundary pollution matters, could regulate them to some degree. Goodsite et al. consider the effect of the RHAP and its cooperative measures, pointing to 'a few notable achievements under this Plan', in particular the establishment of the ASEAN Specialised Meteorological Centre, the 'zero burning policy' and the development of controlled burning guidelines.[108] Jayakumar et al. highlight the relationship between the *Haze Pollution Agreement* and the *Conservation Agreement*, commenting generally that:

> ... it imposes a specific responsibility on the Parties to not pollute the air, soil, fresh water and the marine environment. More specifically, it calls upon Parties to take all appropriate measures towards air quality management. In addition, in the area of forest management, the ASEAN Conservation Agreement has addressed the importance of preventing forest fires and developing plans for reafforestation and afforestation.[109]

The *Haze Pollution Agreement* consists of six Parts and one Annex. The Parts concern General Provisions (Part I); Monitoring, Assessment, Prevention and Response (Part II); Technical Cooperation and Scientific Research (Part III); Institutional Arrangements (Part IV); Procedures (Part V); and Final Clauses (Part VI). It can be said that few concrete obligations are imposed on the parties with provisions worded in general, flexible terms throughout. Jerger for example highlights the extensive use of the words 'endeavor to', 'as far as possible', and 'promote' in the requirements of the treaty; while not uncommon, and found also in the UNECE *Air Pollution Convention*, this is nonetheless indicative of the lack of prescription in the obligations, making the appearance of implementation and compliance inevitably more straightforward for the

---

[106] Ibid, 143.

[107] Convention on Long-Range Transboundary Air Pollution (Geneva, 13 November 1979) 1302 UNTS 217, in force 16 March 1983; see Jerger above n. 95, 39; Simon Marsden and Elizabeth Brandon, *Transboundary Environmental Governance in Asia: Practice and Prospects with the UNECE Environmental Agreements* (Edward Elgar, Cheltenham, UK, Northampton, MA, USA, 2015) Chapter 7, and in relation to Asia, 225–230, where governance of northeast Asia is examined. Note also the role of the Malé Declaration in south Asia, which is mentioned briefly in Chapter 5.

[108] Above n. 96, 171.

[109] Above n. 91, 311, footnotes deleted.

parties.[110] It is important however to be aware that the *regime* to tackle the haze pollution problem comprises more than just the Agreement itself. Litta notes that the activities undertaken by ASEAN before and after its entry into force are all relevant, explaining why this is so as follows:

> The activities before the adoption of the Agreement were clearly helping to formulate the concrete provisions of it. The activities that have been undertaken after the Agreement's adoption either result directly from the Agreement's provisions or can be seen as indirect consequences of it. Thus, I regard the haze regime as more than just one Agreement. It is a multitude of activities of ASEAN with a legally binding treaty at its core.[111]

The institutional arrangements under the Agreement are provided for in the relationship between the ASEAN Coordinating Centre for Transboundary Haze Pollution Control (Article 5(1), a Committee comprised of representatives of national authorities of the Parties (Article 5(3), and the Competent Authorities and Focal Points of the Parties (Article 6). These are supplemented by the overall control exercised by the COP to the Agreement (Article 18), and its Secretariat (Article 19). The specific arrangements under which the Centre, Competent Authorities and Focal Points operate, are outlined in the next subsection. In relation to the COP and Secretariat, these arrangements are as follows.

Article 18(1) establishes the COP with ordinary meetings to be held at least yearly in conjunction with ASEAN meetings, with extraordinary meetings as necessary (Article 18(2). The primary role of the COP is to continuously review and analyse implementation of the Agreement, including the consideration of reports from Parties and the Secretariat; the adoption of Protocols (Article 21), amendments and Annexes (Articles 22 and 23); and the establishment of subsidiary bodies (Article 18(3)). The Secretariat is established under Article 19(1), and under Article 19(3) the ASEAN Secretariat serves as the Secretariat to the Agreement. Its functions include administering the COP meetings and those of other bodies under the Agreement; sending notifications, reports and other information to Parties; considering information received and consulting with Parties; and ensuring coordination with international bodies (Article 19(2)).

---

110 Above n. 95, 40–41.
111 Above n. 94, 108.

Jerger explains the arrangements that are in place as a 'managerial model' to address the issue.[112] The purpose and advantage of this is that it enables compliance to increase over time, in response to improvements in technology, science and politics. It contrasts with the 'traditional model' that takes a coercive command and control approach, and is therefore very suitable to the political situation of the 'ASEAN Way' which was explained at the start of the chapter. Jerger also draws comparisons with the UNECE *Air Pollution Convention*,[113] which takes a similar cooperative approach, relying on self-reporting by the member States. While this may not always produce compliance, it at least enables communication, which, where lack of compliance may be due to issues of capacity rather than deliberate non-compliance, may help address the problem in more practical ways than naming and shaming an under-resourced developing nation.

4.2 Objectives, Principles and Procedures, including Implementation and Compliance

Haze pollution is defined under the Agreement as 'smoke resulting from land and/or forest fire which causes deleterious effects of such a nature as to endanger human health, harm living resources and ecosystems and material property and impair or interfere with amenities and other legitimate uses of the environment' (Article 1(6)). Under Article 2, the primary aim is to prevent, monitor and mitigate transboundary haze pollution arising from land and forest fires through 'concerted national efforts and intensified regional and international co-operation'. Principles underlying action are outlined in Article 3, which relate to international norms such as state responsibility and the no harm rule (Article 3(1)); cooperation (Article 3(2)); precaution (Article 3(3)); sustainable development (Article 3(4)); and participation (Article 3(5)).

Specifically, Article 4(1) requires Parties to cooperate in developing and implementing measures to minimize and monitor transboundary haze pollution. These include control of the sources of fires; development of monitoring, assessment and early warning systems; exchange of information and technology; and provision of mutual assistance. Article 4(2) requires Parties to respond promptly to a request for relevant information

---

112 Above n. 95, 38.

113 Above notes 95 and 107; see also A Byrne, 'The 1979 Convention on Long-Range Transboundary Air Pollution: Assessing its Effectiveness as a Multilateral Environmental Regime after 35 Years' (2015) 4(1) *Transnational Environmental Law* 37–67.

made by any state that may be affected by transboundary haze pollution, so as to minimize any adverse effects. Article 4(3) states that legal, administrative and other appropriate measures must be taken by Parties to implement their obligations.

Similarly to the response to biodiversity conservation under the related ASEAN *Conservation Agreement* and *Centre for Biodiversity Agreement*, the *Haze Pollution Agreement* also establishes a Centre to address the issue under Article 5. This is known as the ASEAN Coordinating Centre for Transboundary Haze Pollution Control. In light of the other organization, confusingly known as 'the ASEAN Centre' (Article 5(1)), the purpose is to facilitate cooperation and coordination in managing the *impact of* (my emphasis) land and/or forest fires, and haze pollution in particular. The Centre therefore does not concern itself with preventing the fires in the first place under the Agreement, which is addressed via the obligation in Article 9 (Prevention) – below – and the subregional politics of the 'ASEAN Way'. Instead, the work of the Centre is based on the premise that national authorities will first address the problem when it arises by putting out the fires. Only when an emergency situation is declared by the national authority will the Centre potentially become involved, via a request for assistance (Article 5(2)).

The functions of the Centre are outlined in an Annex to the Agreement. Thirteen functions are given to the Centre, including: establishing and maintaining regular contact with National Monitoring Centres; facilitating cooperation and coordination among Parties to increase preparedness and response; establishing and maintaining information on experts, facilities and donors; and responding to requests for assistance by information sharing and coordination of assistance between Parties. Article 6 also concerns administrative arrangements, and requires Parties to designate Competent Authorities and Focal Points, and for this information to be shared with other Parties and the Coordinating Centre. Article 7(1) furthermore requires Parties to designate bodies as National Monitoring Centres to comply with the monitoring obligation (below). Jerger explains the relationship between the Parties and the Centre as follows:

> The Centre … receives data compiled by parties' national monitoring centres … Having a centralized database allows for more efficient communication and thus decreases the time parties need to spend searching for the data … In addition to quicker access to the data, parties are able to understand the data better because the Centre standardizes reporting formats … Parties must also inform the Centre and other parties of the measures they have in place to implement the Agreement. By reporting on these steps, all parties are reassured that each is making a good faith effort to comply with the

Agreement, which will reinforce the parties' confidence in one another and lead to greater coordination.[114]

The substantive obligations of the Parties are found in Articles 7 (Monitoring), 8 (Assessment), 9 (Prevention) and 10 (Preparedness). The obligation in Article 7(3) is highly significant, requiring that 'Parties, in the event that there are fires, *shall initiate immediate action to control or to put out the fires*' (my emphasis). The realization of this obligation has, however, as indicated in the introduction to this subsection of this chapter, clearly not been achieved. Article 7 (Monitoring) also requires Parties to monitor 'fire prone areas', which are to be defined by national authorities as 'areas where fires are most likely to occur or have a higher tendency to occur' (Article 1(4)); 'land and/or forest fires', which are defined as 'fires such as coal seam fires, peat fires, and plantation fires' (Article 1(7)); environmental conditions conducive to the fires; and haze pollution resulting from them, which is defined as 'smoke resulting from … fire which causes deleterious effects of such a nature as to endanger human health, harm living resources and ecosystems and material property and impair or interfere with amenities and other legitimate uses of the environment' (Article 1(6)).

Article 8 (Assessment) obliges Parties to ensure communication between National Monitoring Centres and the Coordinating Centre, and lays down the procedure for the receipt of communications, its analysis and risk assessment in relation to effects upon health or the natural environment. Article 9 (Prevention) requires Parties to take measures to not only prevent, but also to *control* (my emphasis) activities related to fires that 'may lead to transboundary haze pollution'; this is defined as 'haze pollution whose physical origin is situated wholly or in part within the area under the national jurisdiction of one Member State and which is transported into the area under the jurisdiction of another Member State (Article 1(13)). Seven measures are identified, with the first (Article 9(a)), to develop and implement regulation, programmes and strategies to promote a 'zero burning policy'. This is defined as 'a policy that prohibits open burning but may allow some forms of controlled burning' (Article 1(14)); in turn, 'open burning' is defined as 'any fire, combustion or smouldering that occurs in the open air' (Article 1(9)); and 'controlled burning' 'means any fire, combustion or smouldering that occurs in the open air, which is controlled by national laws, rules, regulations or guidelines and does not cause fire outbreaks and transboundary haze pollution' (Article 1(3)).

---

[114] Above n. 95, 41.

Article 9(b) requires Parties to develop other policies to put an end to actions resulting in fires; Article 9(c) to identify and monitor areas prone to fires; Article 9(d) to strengthen fire management and capability and coordination; Article 9(e) to promote education and awareness of the issue to prevent pollution; Article 9(f) to promote and use indigenous knowledge in prevention and management; and Article 9(g) to ensure that regulation controls open burning and prevents the use of fire in land clearing. This latter practice is closely associated with the removal of forests and other vegetation for palm oil plantations in southeast Asia, which is a major incentive for those who break the law as it currently stands. Article 10 requires Parties to develop strategies and response plans to 'identify, manage and control risks; and prepare standard operating procedures for regional cooperation and national action. Articles 11–15 concern emergency response and the provision of assistance.

In accordance with Article 11, Parties must ensure 'appropriate legislative, administrative and financial measures are taken ... to respond to and mitigate' the effects of fires and consequent pollution. An example of this is the legislation passed by Singapore known as the *Transboundary Haze Pollution Act*. This financially penalizes companies for pollution impacting Singapore but originating from activities beyond the border. The Act may set a precedent for future actions but is likely to meet substantial challenges in implementation.[115]

Under Article 12(1), assistance may be requested from the Centre, other states or international organizations in relation to fires within their own territories. Related provisions use the terminology 'requesting Party' (Article 1(12)), 'receiving Party' (Article 1(11)) and 'assisting Party' (Article 1(1)), to distinguish between those asking for help, and those asked for the help; those obtaining it (who may be the same), and those providing it. Requests are either made directly or coordinated through the Centre, and those to whom they are directed 'shall promptly decide and notify', whether they are able to help and, if so, the extent of such help (Article 12(3)). In reply, Parties offered assistance must respond similarly concerning whether they are able to accept offers of help (Article 12(4)). Information must be made available by the requesting Party to the assisting Party, and consultation take place, to enable assistance to be effectively provided (Article 12(5)). The Centre must subsequently be

115 Janice Ser Huay Lee, David P Bickford, Luis R Carrasco, J Jackson Ewing, Zeehan Jaafar, Lian Pin Koh, Alan Khee Jin Tan and Edward L Webb, 'Toward Clearer Skies: Challenges in Regulating Transboundary Haze in Southeast Asia' (2016) 55(1) *Environmental Science and Policy* 87.

advised by the Parties what resources are potentially available to provide the assistance (Article 12(6)).

The direction and control of assistance is dealt with by Article 13, which provides firstly that either the requesting or receiving Party is to exercise overall management of the assistance within its territory; where personnel are involved it must be made clear under whose operational supervision such persons are acting (Article 13(1)). Local facilities and services must be provided by either the requesting or receiving Party to properly administer the assistance, and the protection of the resources provided by the assisting Party must be ensured (Article 13(2)). Parties providing or receiving assistance must coordinate it within their own territory (Article 13(3)). Article 14 requires exemptions to be provided (from taxation, duties or other charges), and facilities to be provided to those giving assistance; Article 15 in turn requires the transit of personnel and related resources to be enabled by those requesting or receiving help. Related provisions provide for technical cooperation (Article 16) and scientific research (Article 17) to underlie efforts by the Parties to resolve the issue.

Jerger rightly draws attention to the significance of the parties undertaking technical cooperation in the field to, under Article 16(1)(g), '[p]romote the development of markets for the utilisation of biomass and appropriate methods for disposal of agricultural wastes'; and under Article 16(1)(a), '[f]acilitate mobilisation of appropriate resources within and outside the Parties'. He refers to this as the 'biomass' provision, commenting that it: 'recognizes that peatlands, as biomass, are a large cause of the fires that create transboundary haze. By developing markets for biomass, the Agreement incentivizes the harvest and controlled burning of peat by turning it into a product rather than a byproduct'.[116]

Implementation is to be reported on further to the obligation in Article 25, which provides that Parties must send to the Secretariat 'reports on the measures taken for the implementation of this Agreement in such form and at such intervals as determined by the Conference of the Parties'. Compliance is summarized in the extremely brief Article 27 (Settlement of Disputes), which simply states – in accordance with the 'ASEAN Way' – 'Any dispute between Parties as to the interpretation or application of, or compliance with, this Agreement or any protocol thereto, shall be settled amicably by consultation or negotiation'. Litta notes this and comments:

---

116 Above n. 95, 42, footnote deleted.

> There are no provisions in the Agreement about sanctions for non-compliance with the provisions of the Agreement. Automatic compliance is assumed. Compliance will be rewarded with good neighbourhood relations and assistance in crisis situations. The Association has always avoided spelling out sanctions for non-compliance in order to keep a good spirit of cooperation.[117]

As noted above, the effectiveness of the *Haze Pollution Agreement* is also significantly undermined by the lack of participation by Indonesia. Indonesia is the largest source of forest fires within the region, and transboundary haze pollution therefore continues to present a challenge for ASEAN.[118] However, even though it is not a party to the Agreement, Indonesia has taken measures to deal with the situation to some extent. Jerger notes the Indonesian Plan of Action and efforts by an NGO, the Roundtable for Sustainable Palm Oil, as examples. The work of the Sub-Regional Ministerial Steering Committee on Transboundary Haze Pollution is given emphasis, in part because of the benefits that have arisen by way of 'sharing concession maps that show where burning is taking place and holding those who are doing the burning responsible'.[119] Litta is particularly critical, however, noting that:

> The Agreement does not address the root causes of the fires (natural resource management, land tenure conflicts, crony capitalism, mega-development projects, illegal logging in protected areas, and rampant corruption in all levels of government). The Agreement concentrates on technical solutions. Thus, even if Indonesia were to fully comply with the treaty's obligations, these are but immediate and short-term measures, which do not address the long-term fundamental problems of unsound forest management in Indonesia.[120]

Because of this and all the other problems with this Agreement, as seen above, it has therefore been said that: 'As yet, ASEAN has not fully achieved effective environmental governance for sustainable development

---

[117] Above n. 94, 116.

[118] Aside from 2015, for example, Singapore also suffered one of its worst haze episode in 2013; see National Environment Agency, 'Factsheet on Transboundary Haze Pollution' (Government of Singapore, 2014) 1. For another recent evaluation, see: Laely Nurhidayah, Zada Lipman and Shawkat Alam, 'Regional Environmental Governance: An Evaluation of the ASEAN Legal Framework for Addressing Transboundary Haze Pollution' (2014) 15(1) *Australian Journal of Asian Law* 1–17.

[119] Above n. 95, 43.

[120] Above n. 94, 117.

but is working towards it. [It] is not very effective in solving critical transboundary environmental issues involving state responsibility'.[121]

5. CHINA'S ROLE IN SOUTHEAST ASIA

China's connections to southeast Asia are considerable. Whether in relation to the environment, economics, social or security relations, it has a long history of engagement with the subregion that continues to this day. ASEAN-China Dialogue Relations began when the Chinese Foreign Minister attended the opening session of the Twenty-Fourth ASEAN Ministerial Meeting in July 1991 in Kuala Lumpur, in which he expressed China's keen interest to cooperate with ASEAN for mutual benefit. Subsequently, China was accorded full 'Dialogue Partner' status at the Twenty-Ninth ASEAN Ministerial Meeting in July 1996 in Jakarta. The relationship was enhanced with the signing of a Joint Declaration at the Seventh ASEAN-China Summit in October 2003 in Bali; China appointed its first resident Ambassador to ASEAN and established its Mission to ASEAN in Jakarta in September 2012. ASEAN and China are currently implementing the Plan of Action 2016–2020 to implement the Joint Declaration, which was adopted by the Foreign Ministers of ASEAN and China in November 2015.[122] Environmental cooperation between ASEAN and China is noted to include the following:

> [T]he ASEAN-China Environmental Protection Cooperation Strategy 2009–2015 was adopted by the Environment Ministers of ASEAN Member States and China at the 8th ASEAN Plus Three Environment Ministers Meeting in October 2009 to serve as a guiding document for enhancing ASEAN-China cooperation in common priority areas of environmental protection up to 2015. Subsequently, the China-ASEAN Environmental Cooperation Centre (CAEC) was established in Beijing on 17 March 2010 and officially inaugurated on 24 May 2011 to implement the ASEAN-China Environmental Protection Cooperation Strategy 2009–2015. Activities and programmes implemented under the Cooperation Strategy include ASEAN-China Environmental Cooperation Forum, China-ASEAN Green Envoys Program, Biodiversity and Ecological Protection Cooperation, Environmental Technology and Industry Cooperation, and Joint Research.[123]

---

[121] Kheng-Lian and Karim, above n. 8, 463.

[122] See: <http://www.asean.org/?static_post=overview-asean-china-dialogue-relations> accessed 24/5/16.

[123] Ibid, para 33.

In recent times, as China has developed rapidly at home and pursued an expansionist foreign policy, relationships between China and ASEAN have however been less than harmonious. The reason for the disagreements that have arisen and strengthened have been squarely linked with two key issues of concern to southeast Asian nations, particularly ASEAN and the countries of the MRC that overlap. While the focus of this book is upon environmental regimes, it was seen in Chapter 2 how these may be compromised by conflict, such as impacts upon the Arabian Gulf during war; and in Chapter 3 how resource exploitation without due regard to environmental effects can have long-term consequences, such as in the Caspian and Aral Seas. If the same is not to occur – or continue to occur – in the South China Sea and Mekong River Basin therefore, there is a need for these issues to be squarely dealt with.

The first issue is therefore the territorial disputes and linked resource conflicts in the South China Sea, which, as discussed in Chapter 3, is the route of the Maritime Silk Road. These disputes and conflicts have seen China enhance and enforce its claims in the South China Sea by the construction of artificial islands, and the establishment of other offshore structures for hydrocarbon exploration and development; they have also led to the denial of and opposition to safe passage for ships and overflight of aircraft, which have been vigorously resisted by ASEAN nations[124] and other states. So much so in fact that the Philippines brought a case before the Permanent Court of Arbitration (PCA) in The Hague,[125] which in July 2016 was comprehensively determined in favour of the Philippines.[126]

Despite membership of the Law of the Sea Convention (LOSC), which was the primary treaty upon which the claim was based and which has the approach that settlement of disputes must be multilateral,[127] China chose to reject this in favour of attempted bilateral settlements with

---

[124] See for example, Reuters, 'ASEAN Expresses Serious Concern Over South China Sea Developments', *EJ Insight*, 29 February 2016; Jiangtao Shi, 'South China Sea Tensions to Dominate Sino-Asean Talks in Singapore', *South China Morning Post*, 27 April 2016.

[125] For background, see: Jia Talmon, *The South China Sea Arbitration* (Hart, Oxford, 2014).

[126] <https://www.pcacases.com/web/view/7> accessed 24/5/16.

[127] Donald R Rothwell and Tim Stephens, *The International Law of the Sea* (2nd edition, Hart, Oxford 2016).

disputants.[128] China therefore did not appoint an agent to the PCA, and in a 'Note Verbale' on 1 August 2013, reiterated 'its position that it does not accept the arbitration initiated by the Philippines'.[129] Although China may have a legal argument – which is unsupported by the international community[130] – the political reason for its claims is no doubt because as the most powerful nation in the subregion, it is more likely to achieve its goals bilaterally; in contrast, multilateralism favours smaller nations like the Philippines.

As to the legal outcome, Ford believes it is unlikely that this will bring about an end to the dispute;[131] China is also likely to expand its land reclamation and construction activities, exacerbating the issue further.[132] However, its actions may prove counter-productive to its efforts to bolster national security, if southeast Asian states are pushed towards the US for their own security, especially if they have access to other means of financial support.[133] Indeed China may find itself isolated based on its own actions and rhetoric,[134] which have recently seen the Chinese Foreign Minister uncomfortable in the international spotlight.[135] Despite

---

[128] See also for example its attempts to resolve conflicting claims via informal means: Zhen Liu, 'China Floats Plans for International Maritime Legal Centre Amid Territorial Disputes', *South China Morning Post*, 13 March 2016.

[129] Above n. 126. See John Ford, '*Philippines v. China* Won't End South China Sea Disputes', *The Diplomat*, 17 March 2016, where Ford comments: 'In its position paper, China argued that the Court had no jurisdiction over the nine-dash line because it is based on underlying claims to territory (namely, China's claims to the island chains of the South China Sea) and the Court is expressly prohibited from deciding territorial disputes. Essentially, China argued that the Court could not rule on the legality of the nine-dash line because it did not have the power to decide the underlying dispute over control of the Spratly Islands'.

[130] This has been explained by the former first Chief Executive of the Hong Kong Special Administrative Region of the People's Republic of China. See Tung Chee-hwa, 'Right Side of History', *South China Morning Post*, 26 May 2016.

[131] Ford, above n. 129.

[132] Julian Borger, 'South China Sea Fears Grow Before Tribunal Rules on Disputed Islands', *The Guardian*, 29 May 2016.

[133] See Teddy Ng, 'Beijing "May Need Rethink" as Vietnam Moves Closer to US', *South China Morning Post*, 30 May 2016.

[134] Agence France-Presse, 'South China Sea: US Warns Beijing Against Building "Great Wall of Self-Isolation"', *The Guardian*, 4 June 2016.

[135] Ashifa Kassam and Tom Phillips, 'Chinese Minister Vents Anger When Canadian Reporter Asks About Human Rights', *The Guardian*, 2 June 2016.

intervention by the Chinese President, this is unlikely to break the deadlock in the positions of the states involved, without a change of policy.[136]

The second issue is the impact upon the Mekong River Basin by the extensive dam-building activities discussed in earlier sections, and which are also pursued by China. Mekong Basin development is one of the 11 priority areas of cooperation between China and ASEAN, so it is not without support from ASEAN. However, while China is not alone in this, as noted above, with Vietnam, Thailand and Laos all involved also, it holds the key position as the state where the origin of the Mekong (and other) rivers begin their journey to the sea. Despite concerns raised by the MRC, which have slowed development on the Mekong in recent years, China has advocated strongly against this position, and in favour of more dam building in the subregion.[137]

Furthermore, China has benefitted from its powerful position to also influence the situation in the South China Sea.[138] While it has used its upstream status on the Mekong to provide aid (as a carrot), as well as restrictions in water flows (as a stick), China's primary and traditional opponent in Asia – Japan – while not being involved in the Mekong River Basin geographically or politically as a member of the MRC, has attempted to counter Chinese influence by providing aid of its own.[139] As well as balancing China's role within Asia, the reason for Japan's involvement in the Mekong is related to the ADB, which historically has been a means for Japan to exert its own influence. Japan now has a

---

[136] Nectar Gan and Zhen Liu, 'Xi Jinping Holds Out His Hand to Washington as China-US Talks Begin', *South China Morning Post*, 6 June 2016.

[137] See Oliver Hensengerth, 'Water Governance in the Mekong Basin: Scalar Tradeoffs, Transnational Norms and Chinese Hydropower Investment', in P Nyíri and D Tan (eds), *How Chinese Engagements are Changing Southeast Asia: People, Money, Ideas and Their Effects* (University of Washington Press, Seattle, WA, 2015); Beth Walker, 'China Gives Green Light to New Era of Mega-Dams', in China Dialogue, *Forging a New Course for the Mekong* (China Dialogue, 2014) 4–5; Mekong River Commission, 'China Commits to Continue and Enhance Cooperation with the MRC', *News*, 7 November 2014; and Zhen Liu, 'China Hopes Mekong Nations Summit Will Revive Infrastructure Schemes amid Criticism of Huge Dam Projects on Upper Reaches of River', *South China Morning Post*, 22 March 2016.

[138] Zhen Liu, 'China Pledges Billions to Mekong River Countries in Bid to Boost Influence and Repair Reputation Amid Tensions in South China Sea', *South China Morning Post*, 24 March 2016.

[139] Reuters, 'Japan's Mekong Aid May Counter China', *South China Morning Post*, 3 May 2016.

competitor to the ADB in the China-led Asian Infrastructure Investment Bank (AIIB), as noted earlier and analysed in Chapter 5; this needs to be very carefully managed as the AIIB also has extensive international support, including from a number of western states, such as the UK and Australia.

As with the South China Sea dispute, there is however much to be done in the Mekong subregion if China is to build trust with its neighbours.[140] A change in energy strategy may be needed[141] and greater attention given to those displaced by mega-dams, including the many Indigenous people that reside in affected areas.[142] The benefits of cooperation in the Mekong subregion have been argued as essential to the promotion of 'transboundary cooperation in managing water resources … to bring long-term peace and stability to this part of Southeast Asia … [and] to prevent military conflict and environmental disasters'.[143] Waslekar also draws comparisons with cooperation in central Asia, noting that 'If China, along with Russia, can convince Central Asian leaders to arrive at a pragmatic agreement, it will not only save the environment but also prevent conflict'.[144] As the primary upstream state and dialogue partner within the MRC, and as a dialogue partner within ASEAN, China must – rather than provoke conflict, division and opposition among the ASEAN and MRC states – therefore take a more proactive cooperative role if it wishes to assert itself in the southeast Asian subregion.

6. CONCLUSIONS

Like central Asia, southeast Asia is home to two established environmental regimes, framed by the *Mekong Agreement* and *Haze Pollution*

---

[140] See: Darrin Magee, 'China Fails to Build Trust with Mekong Neighbours', in China Dialogue, *Forging a New Course for the Mekong* (China Dialogue, 2014) 8–10; and Catherine Wong, 'Water Releases "Won't Wash Away Mistrust" Over Mekong', *South China Morning Post*, 18 April 2016.

[141] Brian Eyler, 'China Needs to Change its Energy Strategy in the Mekong Region' in China Dialogue, *Forging a New Course for the Mekong* (China Dialogue, 2014) 14–16.

[142] Si Meng, 'The Fate of People Displaced by the Mekong Dams', in China Dialogue, *Forging a New Course for the Mekong* (China Dialogue, 2014) 25–27.

[143] Sundeep Waslekar, 'Asia's Water Can be a Source of Harmony', *South China Morning Post*, 11 January 2016.

[144] Ibid.

Agreement, with another – at least with regard to the governing *Conservation Agreement*, not in force. In the same way that the *Sustainable Development Convention* in central Asia is not in force, it is arguable that the breadth of the provisions in the *Conservation Agreement* are also considered too forward-looking for the ASEAN states to contemplate, and also too much of a compromise for the advancement of economic development; despite this, they may have precedent value in relation to the Third Pole, as noted in Chapter 5. Nonetheless, the two regimes that are in force – concerned largely with the protection of water and air respectively – have made modest contributions to the transboundary environmental governance of southeast Asia to date, and are not without potential to assist with problem solving, even if they are lacking in certain significant ways as a means of actually solving the underlying problems.

The *Mekong Agreement* is, first, structurally unsound in a number of respects, as this chapter has shown. Notwithstanding past efforts, and despite a current detailed governance structure for the MRC and operational structure for its Secretariat, neither has been capable of overcoming the entrenched national interests that have resulted in decisions being frequently taken outside these structures. In part, as with the history of Soviet domination of central Asia, so also with the history of conflict and division among the southeast Asian states, interests have become entrenched in the subregion and the environment has frequently been the captive. As Sneddon and Fox commented 'the Mekong became embroiled in Cold War geopolitics as a symbol of cooperation'.[145] This symbolic cooperation has of course not been without implications for environmental protection and traditional livelihoods.

It was only at the end of this conflict in 1991 that efforts at development were resumed; however the focus was squarely upon economic development from hydro-electricity, with environmental protection and protection of community livelihoods dependent upon the river basin clearly being secondary concerns. The adoption of the *Mekong Agreement* was therefore undeniably aimed at harnessing the potential electricity generation from the river, and the role of the institutions established were directed towards this, albeit with the autonomy of the four Member Countries minimally impacted in any tangible way. Recent efforts have achieved some success with moving the agenda away from a water-centred to an integrated approach, including the needs of IWRM. However, it has continued to be difficult to take action that is in the best

[145] Sneddon and Fox, above n. 26.

interests of the river basin, notably action that accords with the hydrological cycles of the watercourse. The comment that Thailand has been inclined to 'overwrite hydrological complexity with geopolitical simplification in order to facilitate autonomy of decision-making',[146] summarizes the nature of the problem for all of the Member Countries. This is made worse by the identified difficulties of international law capturing the needs of environmental management,[147] which was identified as a fundamental reason for the failure of the Aral Sea arrangements in the previous chapter.

There are several other structural problems with the *Mekong Agreement* that have come to light in recent years. The first of these is no less relevant for any of the other regimes analysed in Chapters 2 and 3 – the lack of adequate public participation provision. This has been put down by Dore to the 'operating rules which limit engagement with the wider basin community'.[148] Recommendations for a solid legal basis for transboundary EIA including public participation have been made to partly address this.[149] The second, which relates clearly to the prevalence and dominance of national interests, is the need for effective implementation and compliance. This is true also of the role of ASEAN in other southeast Asian regimes, as Contreras finds: 'Like ASEAN, the MRC has no means by which to push member countries to harmonize their national policies according to the agreements forged'.[150] Ultimately, the progress with the Xayaburi and Don Sahong projects by their proponents illustrates these failings very clearly indeed, as agreements to reform the PNPCA process will be far too late for meaningful participation processes to impact upon decision making in each case.

Turning to the ASEAN *Conservation* and *Biodiversity Agreements*, given the human induced pressures on the globally biodiverse subregion of southeast Asia, there is a need for much greater attention to be placed upon this regime. Leaving aside the reasons for the failure of ASEAN states to ratify the first treaty, there is no question that biodiversity conservation is a key part of the environmental protection efforts required and, despite multilateral efforts, providing subregional context for implementation and compliance should be given a much stronger focus. The drivers for economic growth of land clearance for development are

---

146 Sneddon and Fox, above n. 26, 191.

147 Sneddon and Fox, above n. 26, 191.

148 Dore, above n. 30, 17.

149 See Hirsch et al., above n. 64, 122; and Rieu-Clarke, above n. 74, 157–158 and 164–165.

150 Contreras, above n. 7, 164.

certainly related to the environmental regime that is focused upon in southeast Asia – the *Haze Pollution Agreement*. This is in large part because of its impact upon urban as well as natural areas. Recognizing the connections between the impacts from land clearance and air quality would therefore benefit both regimes. Recent support by Jayakumar et al. for the *Conservation Agreement* suggests this treaty should not be left dormant because it 'integrates all the ingredients of a good environmental agreement such as inter-generational equity, balance between conservation and socio-economic development, and the importance of economic cooperation'.[151] The same is arguably true of the *Sustainable Development Convention* discussed in Chapter 3.

In relation to the *Haze Pollution Agreement* specifically, the governance failings of this – not unlike the dominance of national interests impacting upon the Mekong regime – are firmly related to the need to address the underlying reasons for slash and burn agriculture, and the undeniable dynamics of the 'ASEAN Way'. Notwithstanding its acknowledged efforts to address the problem, Indonesia remains outside the regime and is severely afflicted by what Mushkat terms a 'pervasive fragmentation of the environmental management system'.[152] Yet without both assistance and pressure being applied by ASEAN for effective implementation and compliance, there is also little likelihood of an end to the air pollution problem. There is no real focus on stopping the fires in the first place, at addressing the reasons for the behaviour, and attention is directed towards technical solutions. There are no sanctions for non-compliance, a preference for cooperation, and, in the words of Elliott, 'a general reluctance to interfere in, or to direct in any authoritative way, the environmental practices of member states'.[153]

Finally, in connection with China's role in the subregion, whether in relation to its dialogue partner status in the Mekong or in its relations with ASEAN, this has been shown to unquestionably focus upon China's interests, rather than those of the subregion. The lack of harmony in China-ASEAN and China-MRC relations clearly relates to the exertion of China's dominance at the expense of cooperative interests. Whether this concerns its focus upon hydro-electric dam development on the Upper Mekong at the cost to downstream users, or its aggressive statements of sovereignty in the South China Sea in relation to other claimants, its self-interest is undeniable. The environment is the loser in

151 Jayakumar et al., above n. 91, 311.
152 Mushkat, above n. 103, 145.
153 Elliott, above n. 15, 60.

both circumstances as impacts on water quality and quantity in the Mekong River Basin, and risks of conflict grow and add to other potential effects upon the marine and coastal environments of the South China Sea. The resource demands of an actively developing economy domestically and internationally are the underlying reasons, which without a changed approach are extremely unlikely to prevent further significant deterioration.

5. Linking south and east Asia: the Tumen regime, China and the Third Pole

1. INTRODUCTION

This final chapter analyses environmental governance in the subregions of both south and east Asia. From the preceding chapters, it is clear that Chinese influence in the Asian subregions is growing and will continue to do so. However, ensuring that the environment is effectively protected at the same time is, while extremely challenging, essential given China's ongoing focus on development. This has been seen to be based primarily upon its own needs for resource security, for further economic development at home and abroad, and its expansionist aims and actions to maintain and consolidate its territorial integrity.

In addition to analysing the potential for an environmental regime for the Third Pole, this chapter considers matters of specific relevance to China in east Asia: its external relations with its neighbours in that subregion, particularly in northeast Asia where it is directly involved as a Party to a subregional environmental regime, the *Tumen Programme*.[1] Further to its involvement in international environmental law in multilateral contexts – which is commented upon in connection with implementation and compliance generally, and specifically in connection with the South China Sea dispute – this direct experience of subregional law will be invaluable for its involvement in any future Third Pole regime.

[1] See 1995 Agreement on the Establishment of the Tumen River Area Development Coordination Committee, signed in New York, 6 December 1995; and 1995 Agreement on the Establishment of the Consultative Commission for the Development of the Tumen River Economic Development Area and North-east Asia, signed in New York, 6 December 1995; collectively the 'Tumen Programme'.

Other environmental frameworks are less developed, and include the UNEP-administered *East Asian Seas Programme*;[2] and the collection of arrangements for air pollution control, particularly in northeast Asia, where, in addition to southwest China, most of its other borders with neighbouring states are found.[3] The role of China and the AIIB in providing finance for development activity more broadly will also be focused upon, together with an analysis of the Third Pole and the environmental and energy issues it faces. Distinctly from Chapters 2–4, which are limited to subregional environmental regimes, Chapter 5 addresses issues of relevance to the Third Pole in a slightly broader context.

In south and east Asia combined therefore, there is only one established subregional environmental regime, the *Tumen Programme* in northeast Asia. Because of this, it will briefly highlight the significance of one bilateral water agreement and one MEA: the *Indus Waters Treaty*[4] and the CBD.[5] These are relevant in particular because of the content of both, as evidence of practice under the former between two states traditionally in dispute in the subregion (India and Pakistan), and as an indication of the importance of biodiversity conservation in the megadiverse area of the Himalayas in particular. Reference to the CBD also enables evidence of China's recent compliance with international environmental law to be commented on, which is relevant to any future involvement it may have in a Third Pole regime.

The chapter will first consider the nascent environmental regimes in the south Asian subregion which are also very relevant to the issues to be addressed in any Third Pole regime. While both south and east Asia are the subregions with the weakest development of regime building, south Asia does however have institutional structures to support this. These can be developed further, in particular the South Asian Association for

---

[2] <http://www.unep.org/regionalseas/programmes/unpro/eastasian/default.asp> accessed 8/6/2016. This is also applicable to southeast Asia but is considered in this chapter because of the involvement of China and South Korea.

[3] See Simon Marsden and Elizabeth Brandon, *Transboundary Environmental Governance in Asia: Practice and Prospects with the UNECE Environmental Agreements* (Edward Elgar Publishing, Cheltenham, UK and Northampton, MA, USA, 2015) 225–230.

[4] Indus Waters Treaty between the Government of India and the Government of Pakistan (Karachi, 19 September 1960), 419 UNTS 126, in force 1 January 1961.

[5] Convention on Biological Diversity (Rio de Janiero, 22 May 1992), 31 ILM 822, in force 29 December 1993.

Regional Cooperation (SAARC) and the South Asia Co-operative Environment Programme (SACEP). Together with the role of the major state in south Asia (India), these are explored at length in this chapter because of the importance of subregional leadership to regime building; the limitations found with the MRC and ASEAN in Chapter 4 emphasize this aspect.

As to the undeveloped environmental regimes in south Asia, an *Environmental Cooperation Convention*[6] has recently entered into force, although details are unavailable and no action appears to have been taken to date. However the subject area of both this treaty and the other arrangements is highly relevant in terms of regime building for the Third Pole. The other agreements focus upon water supply issues, notably the aforementioned *Indus Waters Treaty*, or other energy issues, in particular the *Energy Cooperation (Electricity) Agreement*, although this is not in force.[7] Given the disaster-prone nature of this subregion, a *Rapid Response to Natural Disasters Convention* has also been adopted, although again this is not in force.[8] However, both the *Energy* and *Disasters* agreements can readily form the basis for a comprehensive regime for the Third Pole, supported by development of the more rudimentary measures for climate change, which are also discussed here. Of no less relevance to the Third Pole, is the non-UNEP-administered *South Asian Regional Seas Programme*[9] for the marine environment; this is important because it addresses challenges in the delta and coastal zone of the transboundary rivers that originate in the Tibetan Plateau. Finally, the *Malé Declaration* in respect of air pollution[10] has been promulgated, which is also of relevance to the Third Pole given the consequences for global warming from the burning of fossil fuels that result in air

---

6 Convention on Cooperation on Environment (Thimphu, 28–29 April 2010), unreported, in force 23 October 2013.

7 Framework Agreement for Energy Cooperation (Electricity) (Kathmandu, 27 November 2014), unreported, not in force.

8 Agreement on Rapid Response to Natural Disasters (Maldives, 10–11 November 2011), unreported, not in force; at November 2014, five states had ratified.

9 <http://www.unep.org/regionalseas/programmes/nonunep/southasian/default.asp> accessed 8/6/2016.

10 Malé Declaration on Control and Prevention of Air Pollution and Its Likely Transboundary Effects for South Asia, 1998. See: <http://www.rrcap.ait.asia/male/Pages/Agreement.aspx> accessed 28/1/16.

pollution.[11] Plans for the development of a large number of coal-fired power stations are a particular concern.[12]

There are therefore three main sections to this chapter: south Asia, east Asia and the Third Pole. The structure and rationale is as follows. Section 2 will focus upon south Asia, defining the subregion and analysing existing institutions and regimes. Unlike east Asia, the presence of the institutions and the fact that they are responsible for most of the arrangements in the subregion are indicative of where any future environmental governance for the Third Pole – potentially as supported by China – may lie. Section 3 will analyse east Asia, defining the subregion and China's role, and analysing in particular the *Tumen Programme*, which was identified in Chapter 1 as the ninth and final existing subregional environmental regime; this will be considered in the same way as the others in Chapters 2–4. The *Tumen Programme* is therefore the most formalized regime in either south or east Asia; importantly, it is also the only treaty-based subregional regime analysed in this book in which China has a formal role as a Party.

Section 4 will focus on the Third Pole, providing an overview of geography, political constraints to governance and examples of major issues of environmental concern. The focus will be upon water management and biodiversity protection, and the application of EIA to major infrastructure development as supported by the international lending banks. International cooperation in the Brahmaputra Basin for example, between China, India and Bangladesh, is essential to meet these challenges, which include significant hydro-electric development. The 'New Silk Road' which is also known as the 'Economic Belt' (in the 'One Belt/One Road' scheme as discussed earlier in this book), is not located in this area (as it passes from Xinjiang into central Asia).

However, the establishment of the China-Pakistan Economic Corridor (which also commences in Xinjiang), is certainly relevant to environmental governance of the Third Pole, for which a tenth subregional environmental regime is proposed. Together with the China-led AIIB, the

---

[11] Z Lu, DG Streets, Q Zhang, Q and S Wang, 'A Novel Back-Trajectory Analysis of the Origin of Black Carbon Transported to the Himalayas and Tibetan Plateau During 1996–2010' (2012) 39 *Geophysical Research Letters* LO1809.

[12] Suzanne Goldenberg, 'Plans for Coal-Fired Power in Asia are "Disaster for Planet" Warns World Bank', *The Guardian*, 5 May 2016. Support for this by Australia by approving the export of coal is also of considerable concern. See: <https://www.theguardian.com/environment/carmichael-coalmine> accessed 23/6/16.

implications of this 'Economic Corridor' for the Third Pole will also be evaluated. A concluding section 5 follows, summarizing how Third Pole environmental governance can benefit from existing institutional structures and regimes – however modest and/or untargeted in south and east Asia – together with the other subregional environmental regimes analysed in Chapters 2–4. Recommendations will be made for future subregional environmental governance, setting out suggested directions for future legal and policy research efforts.

2. SOUTH ASIA

2.1 Defining South Asia

South Asia is defined in different ways, but is invariably distinct from southwest and southeast Asia, based on differences in history, culture, religion and, importantly, colonial occupation, which has been a major influence upon the geopolitics of the subregion. It occupies about 5 per cent of the world's land mass, but is home to about 20 per cent of the world's population. This is expected to rise to about 25 per cent by 2025. The subregion includes Afghanistan, Bangladesh, Bhutan, India, Maldives, Nepal, Pakistan and Sri Lanka. The United Nations Geoscheme for southern Asia also includes Iran in this list of nations,[13] which has been considered in Chapter 3. South Asia's topography includes mountains, plateaus, dry-lands, basins and beaches. It varies from the world's highest point, Mount Everest, to coasts, which extend for about 10,000 km from Pakistan to Bangladesh, with wetlands occupying an area of 134,161 km^2. Economically, south Asia is among one of the least integrated regions of the world with regard to trade. The reasons for this are in large part due to geopolitics, which has been expressed by Tripathy as follows:

> There are several, interrelated complexities in South Asia that restrict its regional integration. The acrimonious relations between India and Pakistan, their perpetual conflicts, are a major constraint for integration … Due to its geographical size, economic growth and strategic depth, India is presumed to be a regional hegemon and this affects South Asian integration … Questions of security in South Asia, which are related to cross-border terrorism, also deter the integration process …[14]

---

[13] <http://unstats.un.org/unsd/methods/m49/m49regin.htm#asia> accessed 1/2/16.

[14] Dhananjay Tripathi, 'Is Economic Growth the Required "Catalyst" for South Asian Integration?' UNU-CRIS Working Papers W-2013/12 (United

Despite this, as Gochhayat explains, for India – the most powerful south Asian state – 'Regionalism has remained perhaps the most potent force in Indian politics ...'.[15] This derives from the vast cultural diversity of the country of more than a billion people, in a land of 3,287,263 km^2. There are 28 federal states and seven centrally ruled union territories, and despite over 80 per cent of the population identifying as Hindu, there are also significant other religious groups, including Muslims, Sikhs, Buddhists, Christians and others. This diversity is reflected in India's regions, which are hugely varied, and this has in turn challenged the unification of the state since Independence in 1947, with calls for separate statehood not uncommon. This domestic characteristic is therefore shared with China, and emphasizes the need to cooperate rather than maintain conflict with neighbouring foreign states.

Indian Prime Minister Narendra Modi stressed the need for closer regional cooperation between SAARC countries at an address to the Eighteenth SAARC Summit in Kathmandu in late 2014: 'For India, our vision for the region rests on five pillars – trade, investment, assistance, cooperation in every area, contacts between our people – and, all through seamless connectivity'.[16] As noted below however, despite the efforts of the related organization SACEP, environmental protection has not been a strong feature of this. Significantly also, this subregion is very vulnerable to climate change, which has produced major flooding incidents and related effects upon livelihoods.[17] Although Mirza describes flooding as a 'regular hazard in South Asia', he emphasizes the link with climate change in the following terms:

> The frequency of extreme floods is also apparently on the rise. Bangladesh experienced five large floods from 1987 to 2007. Frequency of increases in flooding is also evident in India and Pakistan. In addition to this, there is a

---

Nations University, Brugge, 2013) 1, 7. See also Wakana Takahashi, 'Environmental Cooperation in South Asia', in Kazu Kato and Wakana Takahashi, *Regional/Subregional Environmental Cooperation in Asia* (IGES, 2001) 51–63.

[15] Artatrana Gochhayat, 'Regionalism and Sub-Regionalism: A Theoretical Framework with Special Reference to India' (2014) 8(1) *African Journal of Political Science and International Relations* 10, 18.

[16] Suhasini Haidah and Damakant Jayshi, 'Modi Talks About "Closer South Asian Integration" at SAARC Summit', *The Hindi*, 26 November 2014.

[17] East Asia is also subject to many of these extreme events, see Rolf Lidskog, 'Extreme Events, Regulatory Style and Regional Environmental Governance', in Jörg Balsiger and Aysun Uyar (eds), *Comparing Regional Environmental Governance in East Asia and Europe: Proceedings* (Research Institute for Humanity and Nature, Kyoto, 2013) 53–63.

> tendency of expansion of flood-prone areas. For example, in western India where droughts are very frequent, floods are also occurring frequently. In July 2005, in Gujarat, floods engulfed a large region and killed 132 people ... Although these events are within the range of climate variability, based on the future climate scenarios from the global climate models (GCMs) scientists are concerned that the frequency, magnitude and extent of flooding may increase in South Asia.[18]

Impacts of climate change have been observed in the form of glacier retreat in the Himalayan region, where in 2009 it was reported by the UN Under-Secretary General and Executive Director of UNEP that approximately 15,000 glaciers will likely shrink from the present total area of 500,000 km^2 to 100,000 km^2 by 2035.[19] In 2014 the outlook had deteriorated even further, with the Intergovernmental Panel on Climate Change identifying transboundary pollution as an intra- and inter-regional issue of significance.[20] Taking India as an example, the environmental consequences of development have been growing rapidly in recent decades. Swain indicates the main reasons for environmental degradation, highlighting the economic circumstances of the region's people as the catalyst of change.

> The region is very vulnerable to environmental problems because of its high population growth rates and density, and its high levels of poverty. Of the region's population, which is expected to reach 1.5 billion by 2020, more than 50 percent live below the poverty line and lack livelihood options. Despite rapid economic growth, particularly in India during recent years, the region has one of the lowest average per capita incomes in the world.[21]

---

[18] Monirul Qader Mirza, 'Climate Change, Flooding in South Asia and Implications' (2011) 11 *Regional Environmental Change* S95, S95, reference deleted.

[19] Achim Steiner, *South Asia Environment Outlook 2009* (UNEP, SAARC, 2009), foreword, p. iii.

[20] Y Hijioka, E Lin, JJ Pereira, RT Corlett, X Cui, GE Insarov, RD Lasco, E Lindgren and A Surjan, 'Asia', in VR Barros, CB Field, DJ Dokken, MD Mastrandrea, KJ Mach, TE Bilir, M Chatterjee, KL Ebi, YO Estrada, RC Genova, B Girma, ES Kissel, AN Levy, S MacCracken, PR Mastrandrea and LL White (eds), *Climate Change 2014: Impacts, Adaptation, and Vulnerability. Part B: Regional Aspects. Contribution of Working Group II to the Fifth Assessment Report of the Intergovernmental Panel on Climate Change* (Cambridge University Press, Cambridge, 2014) 1353.

[21] Ashok Swain, 'South Asia, its Environment and Regional Institutions', in L Elliott and S Breslin (eds), *Comparative Environmental Regionalism* (Routledge, Abingdon, UK, 2011).

Despite these problems, governance has been weak. Water management issues predominate over environmental concerns, especially in the transboundary rivers context. These are exacerbated by the national security concerns of many states in the subregion, and the fact that groundwater supplies are fast depleting, for example in India, Pakistan and Bangladesh. Rivers are dependent upon the summer monsoon and alpine snow melt, which are happening later or at much reduced levels than in previous decades. Price et al. highlight the governance failings that must be addressed for these challenges to be met head on:

> In spite of the shared river systems and interdependencies, South Asian governments have signed few bilateral water agreements and no regional ones. Those transboundary water treaties that do exist face criticism on a number of grounds: for time periods [that are] too short or too long; for the absence of dispute resolution mechanisms; and for the absence of provision for environmental factors or new challenges such as climate change.[22]

2.2 South Asian Institutions

2.2.1 SAARC

SAARC was formed by Charter 1985 by the following states: Bangladesh, Bhutan, India, Maldives, Nepal, Pakistan and Sri Lanka; Afghanistan joined later.[23] Most are in the Third Pole subregion, and have agreed to promote 'peace, stability, amity and progress in the region'.[24] In relation to the issues addressed, the Charter makes no mention of the environment in either the recitals or articles, but is focused above all on 'fostering mutual understanding, good neighbourly relations and meaningful cooperation'.[25] However, in 2014, at the Eighteenth SAARC Summit Declaration in Kathmandu, the heads of government expressed concern in relation to disasters and climate change,[26] and 'directed the relevant bodies/mechanisms for effective implementation of SAARC

---

[22] Gareth Price, Rafay Alam, Sarah Hasan, Fahd Humayun, Mohammad Humayun Kabir, Chandni Singh Karki, Sonali Mittra, Tasmia Saad, Mehr Saleem, Samir Saleem, Pragya Ratna Shakya, Christopher Snow and Shayasta Tuladhar, *Attitudes to Water in South Asia* (Chatham House, London, 2014) viii.

[23] Charter of the South Asian Association for Regional Cooperation (Dhaka, 18 December 1985).

[24] Ibid, recital 1.

[25] Ibid, recital 2.

[26] See subsection below on climate change, and for more information see: <http://saarc-sec.org/areaofcooperation/cat-detail.php?cat_id=54> accessed 5/11/15.

Agreement on Rapid Response to Natural Disasters, SAARC Convention on Cooperation on Environment and Thimphu Statement on Climate Change ...';[27] 'renewed their commitment to substantially enhance regional connectivity ... through building and upgrading roads, railways, waterways infrastructure, energy grids, communications and air links ...'; 'emphasized the need for linking South Asia with contiguous regions, including Central Asia ...';[28] and 'directed the relevant SAARC bodies and mechanisms to identify regional and subregional projects in the area of power generation, transmission and power trade, including hydropower, natural gas, solar, wind and bio-fuel, and implement them with high priority with a view to meeting the increasing demand for power in the region'.[29]

The SAARC Charter describes the institution as an *organization*,[30] and the fact that it has entered into numerous agreements in relation to matters such as trafficking, terrorism, narcotics, criminal matters, child welfare and the environment indicate that it may in fact have legal personality.[31] SAARC has in recent years expanded to include observers in its governance framework; at the Delhi 2007 summit for example, six countries – China, the USA, Mauritius, Japan, South Korea, Iran – plus the EU, joined;[32] this trend has been followed since by the admission of

---

[27] Kathmandu Declaration, Eighteenth SAARC Summit, Kathmandu, Nepal, 26–27 November 2014, para 14 (environment).

[28] Ibid, para 8 (connectivity).

[29] Ibid, para 9 (energy). Note that nuclear power development is also a major part of this; see Agence France-Presse, 'Russia to Build 10 More Nuclear Reactors in India', *South China Morning Post*, 12 December 2014.

[30] SAARC Charter, above n. 23.

[31] The ICJ in the Reparation case established the parameters for this. See: *Advisory Opinion: Reparation for Injuries Suffered in the Service of the United Nations*, 11 April 1949, declaration of Judge Winiarski, available at <http://www.icj-cij.org/docket/index.php?p1=3&p2=4&case=4&code=isun&p3=4> accessed 26/6/16. This has been considered extensively since in relation to the Arctic Council; see Sébastien Duyck, *Legal Issues Related to Options for a Regional-Seas-Type Arrangement for the Arctic Ocean* (Northern Institute for Environmental and Minority Law) 3, and the examples given of other institutions (including Asia Pacific Economic Cooperation). See also Eric C Ip, 'The Power of International Legal Personality in Regional Integration', United Nations University-Comparative Regional Integration Studies Working Papers, W-2010/4, which also considers ASEAN.

[32] Suk Deo Muni and Rajshree Jetly, 'SAARC: The Changing Dimensions', United Nations University-Comparative Regional Integration Studies Working Papers, W-2008/8: 20.

additional observers.[33] The inclusion of the USA and China (with sovereignty over Tibet) is particularly significant. Muni and Jetley furthermore believe there is potential for China to become a full member in the future, supported by Pakistan, Nepal and Bangladesh, among others.[34] While this is yet to happen, it would have a major influence on the potential future governance of the Third Pole.

2.2.2 SACEP

SAARC was preceded by SACEP in 1982, emphasizing the importance of sustainable development and cooperative action to address environmental degradation and other issues. SAARC has the same members as SACEP, and in July 2004 signed a Memorandum of Understanding (MOU) with it.[35] Because of its membership and objectives, SACEP acknowledges that it is 'an appropriate forum for action on transboundary environmental issues'. The Hindu Kush Himalayan belt is recognized by SACEP for its biodiversity, comprising as it does 10 per cent of the world's flora. SACEP acknowledges however that the region is prone to natural disasters; it was the site of over 60 per cent of the world's disaster-related deaths in the 1990s. Apart from efforts to address natural disasters, including recent earthquakes in the subregion,[36] the majority of SACEP's work has so far gone into coastal and marine issues, rather than addressing specifically Third Pole concerns.

SACEP ambitiously describes itself as an inter-governmental organization;[37] however other than memoranda of understanding and declarations, it does not appear to have concluded any international agreements. Significantly, distinct from most of the other regional seas,[38] there is no agreement for the Indian Ocean more broadly, which may experience

---

[33] See Kathmandu Declaration, above n. 27.

[34] Ibid, 21. India is also warming to China; see author unknown, 'US$10 Bln in Deals Signed as Modi Visits Beijing', *Hong Kong Economic Journal*, 15 May 2015.

[35] Memorandum of Understanding on Cooperation in the Field of the Environment between the South Asian Association for Regional Cooperation and the South Asia Cooperative Environment Programme. The strengthening of the SAARC/SACEP partnership was underscored by the Kathmandu Declaration.

[36] Jon Boone, 'Afghanistan and Pakistan Earthquake Death Toll Will Rise, Say Officials', *The Guardian*, 27 October 2015.

[37] See also Colombo Declaration on the South Asia Cooperative Environment Programme, 25 February 1981.

[38] In particular the Regional Convention for Cooperation on the Protection of the Marine Environment from Pollution (Kuwait, 24 April 1978) in force on 1 July 1979; and Regional Convention for the Conservation of the Red Sea

greater pressures in connection with China's 'Maritime Silk Road' aspirations. Organizational collaboration with other institutions is an essential component of the work of both SAARC and SACEP, both separately and indeed between them.[39] Article 1(h) sets out the objective of the *SAARC Charter*: 'to cooperate with international and regional organisations with similar aims and purposes';[40] similarly, the Mission of SACEP is 'to work closely with all national, regional, and international institutions, governmental and nongovernmental, as well as experts and groups engaged in such co-operation and conservation efforts',[41] and this is demonstrated in its close relationship with SAARC.

2.3 South Asian Regimes

As indicated above, south Asia is the only Asian subregion which currently lacks a transboundary environmental governance framework supported by the nations of the subregion. In contrast, southwest, central and southeast Asia all have at least one regime coordinated under a treaty and related instruments, and east Asia has one also, as will be seen in the next section. However, there are various governance arrangements in place for the Indian subcontinent: a few bilateral treaties, notably the *Indus Waters Treaty*; the three SAARC Agreements, the *Environmental Cooperation Convention* ('Environment Agreement'), *Rapid Response to Natural Disasters Convention* ('Natural Disasters Agreement') and *Energy Cooperation (Electricity) Agreement* ('Energy Agreement'); also the *Malé Declaration* on air pollution, a collection of arrangements for climate change, and the *South Asian Regional Seas Programme*.

The first (the *Indus Waters Treaty*) regulates via a bilateral treaty the transboundary waters between Pakistan and India, and is framed by an international agreement for a transboundary river that is over fifty years old. Although bilateral arrangements – and MEAs – are not the focus of this book, this treaty is however noteworthy because of its long history of cooperation between two states, which have a record of poor geopolitical relations, and because it clearly has a focus on the Third Pole, which depends on numerous transboundary watercourses.

---

and Gulf of Aden (Jeddah, 14 February 1982), ECOLEX B7 p. 982:13, in force 20 August 1985 (see Chapter 2).

[39] Memorandum of Understanding, above n. 35.

[40] SAARC Charter, above n. 23.

[41] See above n. 37.

The second and third treaties, the *Environment* and *Natural Disaster Agreements* are not well developed, and the fourth, the *Energy Agreement*, does not have an environment protection focus. However, these agreements are nonetheless significant because they illustrate the potential of using existing institutions to develop future transboundary environmental governance in the Third Pole subregion crossing south and east Asia.

The fifth, the *Malé Declaration* on air pollution, sets the framework for modest action in this area, which is related to the sixth, the climate change initiatives of SAARC and SACEP. Finally, the seventh arrangement is the *South Asian Seas Programme*, which has relevance to the Third Pole as the watershed of many transboundary rivers in the coastal zone. Subsections below provide a brief overview of each of these arrangements.

2.3.1 *Indus Waters Treaty*

This is a half century old international treaty for the transboundary river between India and Pakistan. India shares a number of transboundary watercourses with its neighbours, in particular China, Nepal, Bangladesh and Pakistan. Reflecting the traditional tensions with its largest neighbour, there are no watercourse agreements with China, despite concerns by India about the use of waters in Tibet by China which currently flow into India. Of particular significance is the Brahmaputra, which is focused upon in considering developing Third Pole environmental governance below.[42] With respect to Nepal, India shares four key rivers with it: the Kosi (or Koshi),[43] Gandaki, Karnali and Mahakali, with an agreement concerning the latter in force.[44] Despite tensions between India and Nepal concerning the other rivers, the Mahakali Agreement has been considered 'a more balanced agreement and included a flood

---

42 See Isabel Hilton (ed), *Brahmaputra: Towards Unity* (undated), available at: <https://www.thethirdpole.net/brahmaputra/> accessed 17/6/16: and generally A Earle, A Jagerskog and J Öjendal (eds), *Transboundary Water Management: Principles and Practice* (Earthscan, London, 2010).

43 See Hugh King and Laurence Smith, 'Many Rivers to Cross: Evaluating the Benefits and Limitations of Strategic Environmental Assessment for the Koshi River Basin' (2016) 18(2) *Journal of Environmental Assessment Policy and Management* 1601002. And generally in relation to assessment of proposals for hydro-electricity, see Alexander Erlewein, 'Disappearing Rivers – The Limits of Environmental Assessment for Hydropower in India' (2013) 43 *Environmental Impact Assessment Review* 135–143.

44 1996 Mahakali Agreement. See: <http://www.unep.org/dams/documents/ell.asp?story_id=123> accessed 16/6/16.

forecasting and warning system'.[45] With respect to Bangladesh, India shares about fifty rivers with its northeastern neighbour (including the mighty Brahmaputra), and pressures with respect to the Farakka Barage have been negotiated in another 1996 treaty.[46]

The most significant river India shares with its neighbours is the Indus, and avoidance of conflict with Pakistan – the other riparian state – was the reason for the *Indus Waters Treaty*. The Indus Valley has been significant for over 5000 years, with water vital to the civilization that developed around the watercourse.[47] In more recent times, irrigation works were the most extensive in the world, with projects developed prior to the Independence of India and Pakistan under the political authority of the British Raj.[48] The Treaty consists of a preamble, 12 articles and eight annexes. The Preamble indicates that the Treaty is concerned with the rights and obligations concerning the 'satisfactory utilisation of the waters of the Indus system of rivers ...', making provision also for the settlement of issues in relation to the interpretation and application of the Treaty.

The obligations are divided between those for the 'Eastern Rivers' (Article II, the Sutlej, Beas and Ravi) and the 'Western Rivers' (Article III, the Indus, Jhelum, Chenab); other obligations apply to both (Article IV). Generally speaking, the waters of the Eastern Rivers are available for the unrestricted use of India, and the waters of the Western Rivers for Pakistan. However, there are numerous exceptions to both in accordance with Articles II, III and IV, and in the applicable Annexes, such as Annex C (agricultural use) and Annex D (generation of hydro-electric power). Detailed obligations are found in Article IV in relation to a range of matters such as irrigation, flood protection, drainage, prevention of pollution and the use of water for industrial purposes.

Other provisions relate to financial matters (Article V), exchange of data (Article VI) and future cooperation (Article VII). Significantly, Article VIII establishes the Permanent Indus Commission, consisting of Commissioners of each state who will together represent their states in

---

45 Philippe Cullet and Joyeeta Gupta, 'India: Evolution of Water Law and Policy', in Joseph W Dellapenna and Joyeeta Gupta (eds), *The Evolution of the Law and Politics of Water* (Springer, 2009) 157, 167.

46 Ibid.

47 Cullet and Gupta, above n. 45, 158.

48 Aaron T Wolf and Joshua T Newton, 'Case Study of Transboundary Dispute Resolution: The Indus Waters Treaty', undated. See: <http://www.transboundarywaters.orst.edu/research/case_studies/Indus_New.htm> accessed 16/6/16.

relation to all matters; the Commission must meet at least once a year. Matters before the Commission include exchanging information and the giving of notice where required. The primary purpose and functions are set out in Article VIII(4), which are the establishment and maintenance of cooperative arrangements to implement the Treaty. These include: study and report on problems related to the development of waters; settling disputes promptly; inspecting the rivers once every five years; and inspecting works and sites as needed. Dispute settlement provisions are found in Article IX, Annex F and Annex G. Annex F adds 'Neutral Experts' to the role of the Commission, and Annex G establishes a Court of Arbitration.[49]

The ability of the *Indus Waters Treaty* (IWT) to meet modern challenges has recently been considered. Sarfraz states that 'the IWT would seem to have become inadequate in its ability to address climatic variability and hydrological uncertainty; environmental flows; and the exigencies of growing demand, driven by significant increases in population'.[50] He emphasizes that the treaty does not consider climate change or environmental flows, and that the population of both countries has increased three-fold in the last fifty years. Much greater pressures are placed upon water resources as a result of global warming, and there is enhanced demand for water for irrigation and hydro-electricity, among other things. Although both bilateral and multilateral frameworks are beyond the scope of this book, Sarfraz contrasts the relevance of the framework with the UN Watercourses Convention (UNWC),[51] and comments:

---

[49] In relation to the potential for Commissions generally to resolve conflicts see Irene B Brooks and Liu Hongxia, 'River Basin Commissions as a Mechanism for Mitigating and Resolving Conflicts' (2006) *China Environment Series* 202–208; and for an example of a dispute between the Parties which this Indus Commission did not resolve, but which went to the Permanent Court of Arbitration, see: Indus Waters Kishenganga Arbitration (Pakistan v India) (Partial Award of 18 February 2013); see: <http://www.pca-cpa.org/showfile.asp?fil_id=2101> accessed 16/6/16. A case report has been published on this by Shashank Kumar, 'The Indus Waters Kishenganga Arbitration (Pakistan v India)' (2013) 17(30) *ASIL Insights*, 13 May 2013, available at: <https://www.asil.org/insights/volume/17/issue/13/indus-waters-kishenganga-arbitration-pakistan-v-india> accessed 16/6/16.

[50] Hamid Sarfraz, 'Revisiting the 1960 Indus Waters Treaty' (2013) 38(2) *Water International* 204, 205.

[51] Convention on the Law of Non-Navigable Uses of International Watercourses (New York, 21 May 1997, by UNGA Res. 51/229), 36 ILM (1997) 700, in force 17 August 2014. For discussion, see: Flavia Rocha Loure and Alistair

> From this analysis of the strengths and weaknesses of the IWT in relation to the UNWC, it appears that the treaty would need to be amended to account for emerging needs and to promote cooperative water management between the two countries, for which the treaty itself provides guidance under Article XII(3).While the UNWC, as noted above, provides a sound basis for such progression, additional aspects must also be taken into account. Joint research on this aspect can in fact pave the way for better understanding of the UNWC within and among both countries.[52]

As indicated earlier, MEAs are not focused upon in this book, nor is international water law, which increasingly has an international environmental law focus. However, this is not to deny their significance. Both the *UN Watercourses Convention* and the *UNECE Water Convention*[53] play significant roles in Asia,[54] and increasingly in the subregion of south Asia known as the Third Pole.[55] Together, they are influencing attitudes to transboundary water cooperation globally[56] and in Asia, including China.[57] In the light of increased pressures on international rivers, many of which originate in the Tibetan Plateau – such as the Mekong River as

---

Rieu-Clarke (eds), *The UN Watercourses Convention in Force: Strengthening International Law for Transboundary Water Management* (Routledge, Abingdon, 2013).

[52] Above n. 50, 213.

[53] Convention on the Protection and Use of Transboundary Watercourses and Lakes (Helsinki, 17 March 1992), 31 ILM (1992) 1312, in force 6 October 1996. For discussion, see: United Nations Economic Commission for Europe, Convention on the Protection and Use of Transboundary Watercourses and Lakes, *The Global Opening of the 1992 UNECE Water Convention* (UNECE, 2013).

[54] Simon Marsden, 'The Helsinki Water Convention: Implementation and Compliance in Asia' (2015) 20(2) *Nordic Environmental Law Journal* 119–129.

[55] Ruby Moynihan and Bjørn-Oliver Magsig, 'The Rising Role of Regional Approaches in International Water Law: Lessons from the UNECE Water Regime and Himalayan Asia for Strengthening Transboundary Water Cooperation' (2014) 23 *Review of European, Comparative and International Environmental Law* 43.

[56] United Nations Economic Commission for Europe/Attila Tanzi, *The Economic Commission for Europe Water Convention and the United Nations Watercourses Convention: An Analysis of their Harmonized Contribution to International Water Law* (United Nations, 2015).

[57] For a comparison of the regimes with reference to China, see Patricia Wouters and Huiping Chen, 'China's "Soft Path" to Transboundary Water Cooperation Examined in the Light of Two UN Global Water Conventions – Exploring the "Chinese Way"' (2014) 22 *The Journal of Water Law* 229; and in relation to state sovereignty, Patricia Wouters, 'The Yin and Yang of International Water Law: China's Transboundary Water Practice and the Changing Contours of

considered in Chapter 4 – it is therefore worth emphasizing their role in influencing the development of subregional environmental governance in Asia, in particular in the Third Pole, as examined below.

2.3.2 *Convention on Cooperation on Environment*

The *Environment Agreement* was signed during the Sixteenth SAARC Summit in 2010, has been ratified by all member states, and entered into force with effect from 23 October 2013. It identifies 19 areas for cooperation in the field of environment and sustainable development 'through exchange of best practices and knowledge, capacity building and transfer of eco-friendly technology in a wide range of areas related to the environment'.[58] It followed the SAARC Environment Action Plan adopted by the Third Meeting of the SAARC Environment Ministers (Malé, 15–16 October 1997). This identified some of the key concerns and established the boundaries and scope for regional cooperation. Since its adoption in 1997, a number of measures have been implemented by the Regional Centres. In addition to these, implementation has been entrusted to a Governing Council comprising the Environment Ministers of member states.

2.3.3 *Agreement on Rapid Response to Natural Disasters*

The *Natural Disasters Agreement* was signed at the Seventeenth Summit (Maldives, 10–11 November 2011) and will come into force once ratified by all members; there are currently five Parties. The setting up and operationalization of the SAARC Natural Disaster Rapid Response Mechanism – as provided for under the Agreement – would institutionalize regional cooperation among member states in the critical area of disaster response. This followed a 'Comprehensive Framework on Disaster Management 2006–2015', which was adopted in 2006 to address the specific needs of disaster risk reduction and management. The Framework is also aligned with the Hyogo Framework of Action (2005–2015). Member states are in the process of preparing their respective National Plans of Action for implementation of the Regional Framework and, thereafter, an Expert Group Meeting will harmonize the national reports and articulate a Regional Plan of Action.

The focus on 'natural' disasters that the title of this agreement suggests is perhaps misplaced. As is clear from recent scientific research and

---

State Sovereignty' (2014) 23(1) *Review of European Community and International Environmental Law* 67.

[58] See: <http://saarc-sec.org/areaofcooperation/cat-detail.php?cat_id=54> accessed 4/11/15.

scholarship, many supposed 'natural' disasters are in fact man-made, whether they are attributed to global warming resulting in unpredictable weather patterns, or geological activity such as earthquakes.[59] The impacts of significant hydro-electricity development, other forms of energy exploration and development, and air pollution from burning the fossil fuels extracted as a result (particularly coal), are all examples which lead increasingly to disasters in south Asia and the need to address them. This is a key part of the analysis of the issues faced by the Third Pole in a subsequent section of this chapter. As with other responses to environmental challenges, disaster management tends therefore to focus upon responses after disasters have taken place, rather than avoidance beforehand. As with war and conflict, such responses are inadequate to meet the major underlying issues; these will continue to grow in number and severity unless and until the causes are effectively addressed.[60]

2.3.4 *Framework Agreement for Energy Cooperation*

At the Sixteenth SAARC Summit held in Thimphu in 2010, a decision was made to enhance energy cooperation. This was intended to facilitate trade and develop efficient conventional and renewable sources, including hydropower. Subsequently, the Seventeenth SAARC Summit, held in Addu City in 2011, emphasized the importance of concluding the *Energy Agreement*. This recognizes the 'common benefits of cross border electricity exchanges', and supports the 'development of efficient conventional and renewable energy sources including hydropower'.[61] The *Energy Agreement* was concluded in Kathmandu in 2014 and contains 20 articles. Two other agreements are intended to be signed later: the Regulation of Passenger and Cargo Vehicular Traffic amongst SAARC Member States, and the SAARC Regional Agreement on Railways.[62]

---

59 See for example, Malcolm Moore, 'Chinese Earthquake May Have Been Man-Made, Say Scientists', *The Telegraph*, 2 February 2009; and Richard A Lovett, 'Man-Made Earthquakes Cause Seismic Rumblings', *Cosmos Magazine*, 24 March 2014.

60 For the international law response in the context of international environmental law, see Tim Stephens, 'Disasters, International Environmental Law and the Anthropocene', in Susan Breau and Katja LH Samuel (eds), *Research Handbook on Disasters and International Law* (Edward Elgar Publishing, Cheltenham, UK and Northampton, MA, USA, 2016).

61 Recitals 2 and 4, SAARC Framework Agreement for Energy Cooperation (Electricity) (Kathmandu, 27 November 2014).

62 Suhasini Haidah and Damakant Jayshi, 'SAARC Leaders Reach Last Minute Energy Deal', *The Hindi*, 27 November 2014.

The objective in Article 2 is to 'enable cross-border trade of electricity on a voluntary basis subject to the laws, rules and regulations of the respective Member States and based on bilateral/trilateral/mutual agreements between the concerned states'. The Agreement lacks environmental protection provisions and is focused on the regulation of the electricity trade between the SAARC member states. Provisions include: promoting competition in the electricity sector (Article 6); planning cross-border grid interconnections 'through studies and sharing technical information' (Article 7); building, operating and maintaining a transmission system across national borders (Article 8); providing non-discriminatory access to transmission systems (Article 12); sharing knowledge and conducting joint research in relation to the electricity sector (Article 14); and making provision for dispute settlement (Article 16). The failure to include environmental protection provisions in the *Energy Agreement* means that conducting EIA for proposed transboundary electricity infrastructure must be carried out in one of three ways: under either specific treaties; in accordance with customary international law; or, if development is to be funded by MDBs/IFIs, further to established environmental safeguards policies.

2.3.5 *Malé Declaration on Control and Prevention of Air Pollution*

On 19 and 20 March 1998, a round-table policy dialogue regarding the rapidly increasing problem of regional air pollution – with a focus on south Asia – was organized at the Asian Institute of Technology, Bangkok, Thailand.[63] The meeting was arranged by the UNEP Regional Resource Center for Asia and the Pacific in collaboration with the Stockholm Environment Institute. It was attended by senior-level environment ministry officials from south Asian countries, analysts, policy-makers and representatives from environmental NGOs. It agreed on the need for action, noting the experience of Europe, and decided to work on a draft declaration; this was approved in principle and submitted to the Seventh Governing Council of SACEP for adoption.

The subsequent meeting, held in Malé in April 1998, adopted the *Malé Declaration.* This emphasized the need for countries to progress or initiate studies and programmes on air pollution in each south Asian country. The first stage required documenting current knowledge and institutional capacity relevant to air pollution issues. Baseline studies were to be developed, with gaps in current knowledge and capacity becoming apparent. National action plans would fill these gaps, creating

---

63 See: <http://www.rrcap.ait.asia/male> accessed 16/6/16.

a scientific basis for the policy process. Implementation of the action plans then necessitated expertise, equipment and information for quantitative monitoring, leading to analysis and policy recommendations for the prevention of air pollution.

The *Malé Declaration* obliged participants to initiate and take forward programmes to do a number of things, including to: 'assess and analyse the origin and causes, nature, extent and effects of local and regional air pollution'; 'develop and/or adopt strategies to prevent and minimise air pollution'; 'set up monitoring arrangements beginning with the study of sulphur and nitrogen and volatile organic compounds'; 'co-operate in building up standardised methodologies to monitor phenomena like acid depositions and analyse their impacts'; 'improve national reporting systems and strengthen scientific and academic effort'; 'draw up and implement national and regional action plans and protocols'; and 'evolve, as appropriate, institutional structures at the national level, including networking'.

The implementation of the *Malé Declaration* was envisaged in phases, bearing in mind its objectives, content and direction. During the last 16 years of implementation, networks of policymakers and stakeholders have been established, including those for monitoring and impact assessment; several impact assessment studies have subsequently been communicated to policymakers and stakeholders, and policy measures initiated to control air pollutants. Phase V (2014–2016) aimed to promote policy measures to control emissions, including 'short-lived climate pollutants'.

2.3.6 Climate change arrangements

Climate change was the theme of the Sixteenth Summit (Thimphu, 28–29 April 2010) which adopted the 'Thimphu Statement on Climate Change'. This outlines initiatives to strengthen and intensify regional co-operation. The Inter-governmental Expert Group on Climate Change is required to monitor, review progress and make recommendations to facilitate its implementation, reporting to the SAARC Environment Ministers. The Fourteenth SAARC Summit (New Delhi, 3–4 April 2007) formed the background to this Statement. This expressed 'deep concern' over global climate change, and called for a climate resilient to development in south Asia.[64]

---

[64] See generally: Andrew Bliss, Regine Hock and Valentina Radić, 'Global Response of Glacier Runoff to Twenty-First Century Climate Change' (2014) 119 *Journal of Geophysical Research: Earth Surface* 717; and in connection with Third Pole security issues, Katherine Morton, 'Climate Change and Security at the Third Pole' (2011) 53(1) *Survival: Global Politics and Strategy* 121.

During the Twenty-ninth session of the SAARC Council of Ministers (New Delhi, 7–8 December 2007), the Council felt that given the area's vulnerabilities, inadequate means and limited capacities, there was a need to ensure rapid social and economic development to make the SAARC subregion climate change resilient. Further to this decision, a Ministerial Meeting on Climate Change was held in Dhaka on 3 July 2008, preceded by an Expert Group Meeting on Climate Change from 1 to 2 July 2008. The Ministerial Meeting adopted the Dhaka Declaration and SAARC Action Plan on Climate Change.

The SAARC Action Plan on Climate Change (2009–2011) identified seven thematic areas of cooperation: adaptation; mitigation; technology transfer; finance and investment; education and awareness; management of impacts and risks; and capacity building for international negotiations. It lists the areas of capacity building for projects; exchange of information on disaster preparedness and extreme events; exchange of meteorological data; capacity building and exchange of information on climate change impacts (for example sea level rise, glacial melting, biodiversity and forestry); and mutual consultation in international negotiation processes as the Priority Action Plan.[65]

2.3.7 *South Asian Seas Programme*

The *South Asian Seas Programme* is part of the global cooperative initiative of the UNEP Regional Seas Programme,[66] although unlike other subregions that are part of this initiative,[67] it is not, currently at least, framed by an international agreement. It therefore contrasts with the examples in southwest Asia (see Chapter 2 on the *Arabian Gulf/Gulf of Oman Convention* and the *Red Sea/Gulf of Aden Convention*).[68] It is however similar to the *East Asian Regional Seas Programme* (below). SACEP is the secretariat of the South Asian Seas Programme.

The South Asian Seas Action Plan was adopted in March 1995, and focuses on ICZM, oil-spill contingency planning, human resource development and the environmental effects of land-based activities. Although

---

65 For more information, see: <http://saarc-sec.org/areaofcooperation/cat-detail.php?cat_id=54> accessed 5/11/15.

66 The Arctic Ocean has been considered as a candidate to be part of this programme: see Duyck, above n. 31.

67 This includes the Caspian Sea, the Red Sea/Gulf of Aden, and the Arabian Gulf – the first of which is regulated by the Framework Convention for the Protection of the Marine Environment of the Caspian Sea (Tehran, 4 November 2003), 44 ILM 1 (2005), in force 12 August 2006.

68 See above n. 38.

there is no regional convention yet, the Action Plan follows existing global environmental and maritime conventions and considers the LOSC as its umbrella treaty.[69] UNEP describes the area covered by the arrangements as follows, which indicates relevance for the Third Pole as a coastal zone watershed in the delta of many transboundary rivers:

> The South Asian Seas (SAS) region can be categorized into two distinct geographical groups. While Maldives and Sri Lanka are island nations, Bangladesh, India and Pakistan are situated on the Asian mainland. The region has some of the largest and [most] biologically rich marine ecosystems, such as the Gulf of Mannar, Atolls of Maldives and Mangroves of Sundarbans. The presence of perennial rivers such as the Brahmaputra, Ganges, Godavari, Indus, Kelani, Magna, etc. have contributed to large networks of backwaters, estuaries, salt marshes and mangroves.[70]

Despite providing habitats for endangered marine turtles, hosting some of the largest coastal lagoons of the world and forming one of the world's premier coral ecosystems, the environmental problems faced are predictable. These are derived from a vast increase in human populations, oil transportation across the Arabian Sea, heavy use of agricultural and industrial chemicals, harmful fishing practices and ill-planned land use. These have resulted in the destruction of significant habitat, driven many wildlife species almost to extinction and compromised human livelihoods, including the potential loss of a member country, the Maldives, from the rising seas resulting from climate change.

Because of the relationship between glacier melt in the mountains of the Third Pole and associated Tibetan Plateau, impacts upon transboundary water courses from this and upstream hydro-electric development in particular, and the delta and coastal zone protected by the *South Asian Seas Programme*, there is a need for a much stronger regulatory framework. The Programme should therefore be framed by a treaty and associated protocols such as the arrangements in southwestern and central Asia for the gulfs and inland seas there. Without this, implementation by domestic law and compliance with mandatory protective laws will be inadequate.

---

69 United Nations Convention on the Law of the Sea (Montego Bay, 10 December 1982), 1833 UNTS 3, in force 16 November 1994.

70 <http://www.unep.org/regionalseas/programmes/nonunep/southasian/> accessed 17/6/16.

3. EAST ASIA

3.1 Defining East Asia and China's Role

The UN Geoscheme includes the following nations in the list for eastern Asia: China, Mongolia, North Korea, South Korea and Japan; Hong Kong and Macau are also included separately in the list (as part of China),[71] but not Taiwan.[72] East Asia is a subregion that is particularly difficult to define; Russia is not included (it is part of eastern Europe under the Geoscheme), although the Russian Far East was previously noted as the most easterly part of Asia geographically.[73] In comparing east Asia with Europe, Balsiger and Uyar comment on what they perceive as 'regional flux through expansion and regional ambiguity through multiple affiliations' in an environmental context.[74] They note the position of China in particular as of relevance to both east and south Asia, which is significantly supportive of the focus and advocacy of this chapter:

---

[71] See above n. 13.

[72] Despite operating as a de facto state since 1971, China's insistence that Taiwan is part of China – and position on the UN Security Council potentially blocking any moves otherwise – means the UN Geoscheme does not recognize its existence. Given the position of Hong Kong and Macau on the Geoscheme however – with much less autonomy than Taiwan, which operates as a functional democracy – this is curious to say the least, but is explained by international politics rather than law. See: Ralph N Clough, 'The Status of Taiwan in the New International Legal Order in the Western Pacific' (1993) 87 *Proceedings of the Annual Meeting (American Society of International Law)*, Challenges to International Governance, 73–77.

[73] For connections between the Russian 'Far East' and northeast Asia, see Takayuki Shiraiwa, 'Amur-Okhotsk Consortium as an Epistemic Community to Conserve the Transboundary Land-Ocean Ecosystem in Northeast Asia and the Russian Far East', in Jörg Balsiger and Aysun Uyar (eds), *Comparing Regional Environmental Governance in East Asia and Europe: Proceedings* (Research Institute for Humanity and Nature, Kyoto, 2013) 29–32. See also discussion of the *Tumen Programme* below. See also Richard Pomfret, *Regionalism in E Asia: Why it has Flourished since 2000 and How Far Will it Go?* (World Scientific, Singapore, 2011), who in considering ASEAN also in this volume, points to further integration potential.

[74] Jörg Balsiger and Aysun Uyar, 'Introduction: Comparing Regional Environmental Governance in East Asia and Europe', in Jörg Balsiger and Aysun Uyar (eds), *Comparing Regional Environmental Governance in East Asia and Europe: Proceedings* (Research Institute for Humanity and Nature, Kyoto, 2013) 1, 2.

> … what constitutes East Asia is important for environmental and[/]or other issues, since complementary, partially overlapping regional orders exist (e.g. Northeast Asia, Greater Asia, Panasia), or because some East Asian members are sometimes also considered part of other subregions (e.g. China as part of South Asia).[75]

As indicated above, the Third Pole connects south and east Asia. China is central to this connection because the Tibetan Plateau, the largest part of the Third Pole, is – according to China and as accepted by most of the international community – legally part of China.[76] China is also central to the environmental governance of the Third Pole because of its economic strength and expansionist policies and plans, particularly for resource exploration and development in the Asian region.[77] China borders 14 states,[78] and, as noted in Chapter 4 especially, it is therefore not surprising that the profile of existing and the potential for new transboundary disputes with its neighbours has been and is high.[79] This is enhanced by the fact that China has the greatest number of international rivers and lakes in the world (over a hundred), with major ones including the Lancang (Mekong), Nu (Salween), Heilong and Tumen. These are divided into contiguous and successive watercourses; the former such as the Heilong and Tumen in northeast China are transboundary rivers, and the latter such as the Lancang (Mekong) and Nu in southwest China, also cross many riparian states.[80]

Environmental harm to these international rivers has mainly resulted from large infrastructure projects such as dams, and pollution incidents. In the case of the former, neighbouring states and NGOs have notably

---

[75] Ibid.

[76] As with Taiwan, the legal status of Tibet is also contested by some. While China claims that Tibet is an integral part of China, the Tibetan Government-in-Exile maintains that Tibet is an independent state under unlawful occupation. See: <http://freetibet.org/about/legal-status-tibet> accessed 20/6/16.

[77] See Tai-Wei Lim, *Oil and Gas in China: The New Energy Superpower's Relations with its Region* (World Scientific, Singapore, 2010).

[78] Russia, North Korea, Mongolia, Tajikistan, Kyrgyzstan, Kazakhstan, Pakistan, Afghanistan, India, Nepal, Bhutan, Myanmar, Laos and Vietnam.

[79] The Tumen River Basin – discussed in the next section – has been identified by UNESCO for example as among the world's 17 river basins which are potential conflict zones for water disputes in the near to long term; See-Won Byun, 'Provincial Impacts of Multilateral Cooperation: The Greater Tumen Initiative and Environmental Protection in Jilin' (2008/2009) 10 *China Environment Series* 163.

[80] HE Daming, Zhao Wenjuan and Feng Yan, 'Research Progress of International Rivers in China' (2004) 14 *Journal of Geographical Sciences* 22.

complained about China's activities in respect of the Lancang River (Mekong);[81] in regard to the latter, the incident on the Songhua (Sungari) River (which joins the Heilong at the Russian border) has recently been noted in the legal literature.[82] China's role as a global environmental citizen continues to receive significant attention. China is commonly viewed as reluctant to engage with customary international law of relevance to these matters,[83] or, as seen in the South China Sea dispute with the Philippines,[84] to participate in multilateral treaty-based solutions to regional issues; at the same time however, it is keen to enhance its profile by emphasizing its ratification of many MEAs.[85]

Given the status of transboundary EIA in international law,[86] and the need for China to build improved relationships with its neighbours based

---

[81] See Francois Molle, Tira Foran and Mira Kakonen (eds), *Contested Landscapes in the Mekong Region: Hydropower, Livelihoods and Governance* (Earthscan, Abingdon, UK, 2009). In relation to both China and India, see Neda A Zawahri and Oliver Hensengerth, 'Domestic Environmental Activists and the Governance of the Ganges and Mekong Rivers in India and China' (2012) 12 *International Environmental Agreements* 269–298. See also generally <www.internationalrivers.org/china> accessed 17/6/16.

[82] Michael Faure and Song Ying (eds), *China and International Environmental Liability: Legal Remedies for Transboundary Pollution* (Edward Elgar, Cheltenham, 2008).

[83] For examples, see Gerald Chan, 'China's Compliance in Global Environmental Affairs' (2004) 45(1) *Asia Pacific Viewpoint* 69; James A Green, *The Persistent Objector Rule in International Law* (Oxford University Press, Oxford, 2016) 234–235, and for a brief definition of the 'persistent objector rule' at 1; and, for further analysis, Joel P Trachtman, 'Persistent Objectors, Cooperation, and the Utility of Customary International Law' (2010) 21 *Duke Journal of Comparative and International Law* 221–233.

[84] Arbitral Tribunal Constituted under Annex VII to the United Nations Convention on the Law of the Sea, (Philippines v China), Award, 12 July 2016 (Permanent Court of Arbitration, 'PCA').

[85] Generally, see: Hanquin Xue, *Chinese Contemporary Perspectives on International Law: History, Culture and International Law* (Brill Nijhoff, Leiden, 2012). For discussion, see for example: Ben Saul, 'China, Sovereignty, Natural Resources and International Law', in Kate Barclay and Graeme Smith (eds), *East Asia's Demand for Energy, Minerals and Food: The International Politics of Resources* (Routledge, London, 2015); Wouters and Chen, above n. 57; and Srini Sitaraman, *State Participation in International Treaty Regimes* (Ashgate, Farnham, UK, 2009).

[86] See *Case Concerning Pulp Mills on the River Uruguay (Argentina v Uruguay) (Judgment)* (General List No 135, 20 April 2010); note that accession to this is now open to all states – United Nations Economic Commission for Europe, 'UNECE Espoo Convention on Environmental Impact Assessment

on the application of transboundary EIA in the Chinese regional context, this is a key contributor to improving environmental governance and sustainability of Asia and its subregions.[87] In relation to transboundary EIA, this has the potential to harmonize environmental standards in cross-border areas, with substantive and procedural matters such as environmental significance, information provision and public participation agreed upon in advance. In the context of regional economic cooperation, which is applicable in both areas discussed, environmental issues can be addressed cooperatively and cross-border problems potentially resolved. Improved political relations are the ultimate advantage of such an approach, with more active engagement between China and other states on matters of environmental concern.[88] This has not been proven in the case of China however, as the dispute with the Philippines has found:

> Despite China's repeated assertions by officials at different levels, that it has undertaken thorough environmental studies, neither the Tribunal, the Tribunal-appointed experts, the Philippines, nor the Philippines' experts have been able

---

Becomes a Global Instrument', Press Release, 27 August 2014; and Convention on Environmental Impact Assessment in a Transboundary Context (Espoo, 25 February 1991) 30 ILM 800, in force 10 September 1997. For commentary, see: Alan Boyle, 'Developments in the International Law of Environmental Impact Assessments and their Relation to the Espoo Convention' (2011) 20(3) *Review of European Community and International Environmental Law* 227–231; Timo Koivurova, 'Transboundary Environmental Impact Assessment in International Law', in Simon Marsden and Timo Koivurova (eds), *Transboundary Environmental Impact Assessment in the European Union: The Espoo Convention and its Kiev Protocol on Strategic Environmental Assessment* (Earthscan, Abingdon, 2011) 23–25; Timo Koivurova, 'Could the Espoo Convention become a Global Regime for Environmental Impact Assessment and Strategic Environmental Assessment?', in Robin Warner and Simon Marsden (eds), *Transboundary Environmental Governance: Inland, Coastal and Marine Perspectives* (Ashgate Publishing, Farnham, UK, 2012) 323–342; Kees Bastmeijer and Timo Koivurova (eds), *Theory and Practice of Transboundary Environmental Impact Assessment* (Martinus Nijhoff, Leiden and Boston, 2008); and Neil Craik, *The International Law of Environmental Impact Assessment: Process, Substance and Integration* (Cambridge University Press, Cambridge, 2008).

[87] Simon Marsden, 'Developing Approaches to Transboundary Environmental Impact Assessment in China: Cooperation through the Greater Tumen Initiative and in the Pearl River Delta Region' (2010) 9(2) *Chinese Journal of International Law* 393–414.

[88] In relation to Third Pole cooperation, see: Christina Larson, '"Third Pole" Glacier Research Gets a Boost From China' (2011) 334 *Science* 1199; and Jane Qiu, 'Glaciologists to Target Third Pole' (2012) 484 *Nature* 19.

to identify any report that would resemble an environmental impact assessment that meets the requirements of Article 206 of the Convention, or indeed under China's own Environmental Impact Assessment Law of 2002.[89]

3.2 East Asian Regimes

3.2.1 The *Tumen Programme*

As with the other subregional regimes examined in Chapters 2–4, the historical background, geographic context and institutional arrangements are important to consider at the outset. The Tumen River is located in the northeast of China; its first 16 km stretch inland from the sea forms the border between Russia and North Korea; further upstream it is the border between China and North Korea.[90] As with its actions in the South China Sea, China defends its right of navigation to – and within – the East Sea, which is more commonly known as the Sea of Japan and is proximate to the Tumen River mouth. While not as well known or contentious as disputes in the South or East China Sea,[91] conflicts here are nonetheless of considerable significance for China and another cause of tension with Japan. The river provides water to the three riparian states of China, Russia and North Korea for agriculture and industrial development, and helps preserve globally significant biodiversity in the Tumen Delta on the Pacific Ocean, of which the East Sea/Sea of Japan is ultimately a part.

Historically, however, the water quality of the river has been and remains poor, being seriously polluted by industrial sewage.[92] The river

---

[89] Above n. 84, para 989. In connection with China's domestic environmental law enforcement, see: Wang Canfa, 'Chinese Environmental Law Enforcement: Current Deficiencies and Suggested Reforms' (2006–2007) 8 *Vermont Journal of Environmental Law* 163. There has been considerable change in arrangements over the last decade, but enforcement remains a major problem. The details are beyond the scope of this book.

[90] See Takayuki Shiraiwa, above n. 73, 29–32; and UNECE Report, Drainage Basins of the Sea of Okhotsk and Sea of Japan, which considers the Amur River Basin and Lake Xingkai as well as the Tumen River Basin. See: <www.unece.org/env/water/blanks/assessment/okhotsk_japan.pdf> accessed 8/6/2016.

[91] For a summary of the East China Sea issues, see <http://www.bbc.com/news/world-asia-pacific-11341139> accessed 21/6/16.

[92] Report on the State of the Environment in Jilin Province in 2006 and Project Document: Feasibility Study on Tumen Water Protection: <http://www.tumenprogramme.org/?info-612-1.html> accessed 8/6/2016. See also Wang Shi-jun, Wang Dan and Yang Xiang-hua, 'Urbanization and its Impacts on Water Environment in Tumen River Basin' (2002) 12 *Chinese Geographical Science* 273–281.

flows through a region that is rich in oil, gas and minerals, with relatively easy access via rail and sea to the markets of the other states of the region, which also include South Korea, Mongolia and Japan.[93] The economic potential of the region was recognized early on and it was decided to establish an intergovernmental organization to assist with the sustainable development of the Greater Tumen Region (GTR). While the environment has been an integral part of the regime from the beginning, overall however it has focused on development; other than as a reason for changes in energy policy, as noted in recent meetings outlined at the end of this section, this has not changed.

In 1995 North Korea, China and Russia entered into an agreement to establish a Coordination Committee for the area's development;[94] supplemented by an agreement to establish a Consultative Commission,[95] which also included South Korea and Mongolia. The Greater Tumen Initiative (GTI), or the *Tumen Programme*, therefore refers to the intergovernmental cooperation mechanism in northeast Asia that was initially constituted by five states: China, North Korea, South Korea, Mongolia and Russia. China's motivation is likely to have been political rather than economic: to balance the threat posed by Japan, to counter Western protectionism and bloc formation, and ensure peace and stability in the Korean peninsula.[96] Throughout its short history the GTI has been supported by the UN Development Programme (UNDP),[97] and was

---

[93] In connection with regionalism generally and the regional governance of northeast Asia with a China, Japan and South Korea focus, see Hidetaka Yoshimatsu, 'Regional Governance and Cooperation in Northeast Asia: The Cases of Environment and IT', RCAPS Working Paper No. 09-9, Ritsumeikan Center for Asia Pacific Studies, Ritsumeikan Asia Pacific University, March 2010.

[94] Above n. 1.

[95] Above n. 1.

[96] See Jean-Marc F Blanchard, 'The Heyday of Beijing's Participation in the Tumen River Area Development Programme, 1990–95: A Political Explanation' (2000) 9 *Journal of Contemporary China* 271–290.

[97] For an explanation of the role of the UNDP, see Jean-Marc F Blanchard, 'Giving the Unrecognized their Due: Regional Actors, International Institutions and Multilateral Economic Cooperation in Northeast Asia' in DW Drezner (ed), *Locating the Proper Authorities: The Interaction of Domestic and International Institutions* (University of Michigan Press, 2003) 49–76.

originally known as the Tumen River Area Development Programme (TRADP), which began in 1992.[98]

The Consultative Commission, which is composed of government representatives of the initially five, then four member states, is the core decision-making institution,[99] and has to date held 16 meetings, which rotate around the member states. The Consultative Commission is supported by the Secretariat, which is based in Beijing. At the Eighth Meeting in Changchun in 2005, the governments agreed to extend the 1995 agreements for a decade and adopted the Changchun Agreement to increase their contributions. It was not clear at the time whether the Memorandum of Understanding (MoU) on the Environment (below) had been formally extended, although the environment was emphasized as a cross-cutting theme, notably with the establishment of the Cooperation Framework on Environment (CFE) in the 2007 Vladivostok Declaration. The CFE was designed to be a formal trans-border mechanism aiming to synchronize efforts of environmental protection in the GTR.

Subsequent meetings have been held in Vladivostok (2007), Ulaanbaatar (2009), Changchun (2010), Pyeongchang (2011), Vladivostok (2012), Ulaanbaatar (2013), Yanbian (2014) and Seoul (2016); at each of these meetings a declaration had been subsequently published. The environmental governance aspects of these are discussed towards the end of this section, concerning implementation. The role of the Coordination Committee is to consult and coordinate in relation to social, legal, environmental and economic issues raised by any of the Parties. Significantly however, the process of transforming the GTI towards becoming an independent legal entity started in 2010, and – in 2016 – is almost complete.[100] Despite these current and potential arrangements, the *Tumen Programme* is hardly recognized among international lawyers,

---

[98] Further historical information can be located in Richard Pomfret, 'The Tumen River Area Development Programme' Winter (1997–1998) *IBRU Boundary and Security Bulletin* 80–88, particularly the experience of conflict and development of trade in the region.

[99] The other core institutions are the Coordination Committee and Secretariat, which serves both the Commission and Committee. The Coordination Committee is focused on development activities and trade and includes representatives of China, Russia and North Korea, whose territories constitute the focus of the Tumen River Economic Development Area. The Secretariat supports and executes the decisions of the Consultative Commission.

[100] See: Seoul Declaration, 28 April 2016, 'We, representatives of the GTI Member States meeting in Seoul on 28 April 2016 on the occasion of the 16th Meeting of the GTI CC, resolve to take actions to expedite GTI's legal transition process', and 'Strategic Review', 1.

which does little to help the relationship with the international relations community of scholars. Koyano for example comments:

> Unfortunately, in general we have not had many success stories concerning academically constructive dialogue between lawyers and experts of international relations. Furthermore, it is often acknowledged that there is no space for lawyers to say anything constructive about transboundary regional cooperation, particularly in Northeast Asia where there is no legally binding multilateral agreement for regional cooperation.[101]

The effectiveness of the objectives, principles and procedures – notably their implementation and compliance – are as essential to the success of the *Tumen Programme*, as to the other subregional environmental agreements considered in this book. The environment has therefore been an area of concern to the GTI states from the start.[102] Article 2(2) of the *Coordination Committee Agreement* states that: 'The Committee shall consult about, and provide coordination for social, legal, environmental and economic issues raised by any Contracting Party related to its sovereign rights'. Sustainable development is cited as an underlying principle in Article 1(1). The *Consultative Commission Agreement* is framed in similar terms, with the environment cited in Articles 2(3) and 2(4), and sustainable development in Articles 1(1) and 2(4). Of far

---

[101] Mari Koyano, 'Comments', in Jörg Balsiger and Aysun Uyar (eds), *Comparing Regional Environmental Governance in East Asia and Europe: Proceedings* (Research Institute for Humanity and Nature, Kyoto, 2013), 93, 93. This is curious for two reasons: first, multilateral agreements by their nature focus on global issues not regional ones, and, second, the Tumen Agreements analysed in this section are – to some extent at least – legally binding arrangements for regional (or subregional given the approach in this book) cooperation.

[102] There is a limited literature on the environmental aspects of the GTI. Note in particular however, Myungjin Kim, 'Environmental Cooperation in Northeast Asia' (2004) 22 (2004) *Impact Assessment and Project Appraisal* 191–203; See-Won Byun, above n. 79, 163; and Marsden, above n. 87 for further commentary on the EIA obligations in particular. For more recent general consideration, see: Won Bae Kim, Yue-man Yeung and Sang-chuel Choe, *Collaborative Regional Development in NE Asia: Towards a Sustainable Regional and Sub-regional Future* (Chinese University of Hong Kong Press, 2011).

greater significance however, is the 1995 MoU on Environmental Principles, signed by all five contracting parties at the same time as the original two agreements.[103]

The MoU affirms the overall goal of the parties: 'to attain environmentally sound and sustainable development of Northeast Asia and the Tumen River Economic Development Area', in accordance with '... (b) the international consensus reached at the United Nations Conference on Environment and Development; (c) the requirements of national laws and regulations and of bilateral and multilateral environmental agreements; and (d) the environmental requirements of the multilateral development banks of which the Contracting Parties are members'. This application of relevant international environmental law, including customary international law, is clear. Provisions requiring EIA as part of international lending arrangements also all have application.[104] The Contracting Parties furthermore 'affirm their intention to cooperate and coordinate with each other to protect and enhance the environment of the Region and to conduct all development activities in the Region in a manner that does not damage the environment of any Contracting Party, of any other state or of areas beyond the limits of national jurisdiction'.

In relation to procedure, the first section of the MoU is entitled 'Environmental Assessment, Mitigation and Management'. Article 1(1) requires the collection and sharing of baseline data, which is to be undertaken nationally and jointly in a coordinated manner. Article 1(3) states that Contracting Parties will prepare and adopt national laws and policies and bilateral and multilateral environmental agreements where 'required to attain environmentally sound and sustainable development of the Region'. Article 1(5) requires EIA for environmentally significant national and transboundary projects, in particular an EIA report and an Environmental Mitigation and Management Plan (EMMP) to deal with

---

[103] Memorandum of Understanding on Environmental Principles Governing the Tumen River Economic Development Area and North East Asia, New York, 1995. The fact that this document is a MoU rather than an agreement between the parties may of course affect the normative value of its provisions.

[104] The latter is not insignificant in this context, see Elizabeth Smith, 'Implementing the Espoo Convention: An International Financial Institution Perspective' and Leonie Schreve, 'The Equator Principles: A Voluntary Approach by Bankers', both in Bastmeijer and Koivurova (eds), *Theory and Practice of Transboundary EIA* (Martinus Nijhoff, Leiden and Boston, 2008), 313–326 and 327–344. See also Salman MA Salman, *The World Bank Policy for Projects on International Waterways: An Historical and Legal Analysis* (The World Bank, Washington DC, 2009), which includes discussion of the policies on disputed areas and EIA, 204–212.

any anticipated effects. Participation by experts in the affected state is a specific obligation, but not public participation.[105]

Article 1(2) mandates the regional environmental assessments of plans, an SEA obligation. The EMMP must again prevent negative effects and enhance the environment; information is to inform decision-making, as it must to be 'based on the results of the regional EA and other relevant data'. Article 1(4) specifies what the EMMP must include, with reference to 11 key matters in addition to general land use management planning. These include: protection of wetlands, fragile coastal areas, forests and sensitive ecosystems; preservation of biodiversity, including threatened or endangered species and their habitats; and the establishment of protected areas.

Article 1(5) contains project specific EIA requirements, both national and transboundary, although, as with Article 1(2), no content requirements as set out. Article 1(5) imposes an obligation on the Contracting Parties on whose territory the proposed project is located to prepare a project-specific EIA report and EMMP to deal with any anticipated 'significant environmental effects'. In contrast to the assessment of plans, the screening requirement for project proposals is hence of environmental significance, although this is not defined further. Article 1(5) (but not Article 1(2) on plan-SEA) further specifies that participation by experts in the affected state is a requirement, although engagement with the public or NGOs is not.

Articles 1(6)–2(5) contain more general environmental obligations which affect those for SEA and EIA.[106] With regard to access to

---

105 For further detail on these obligations, see Marsden, above n. 87.

106 There are numerous examples in a Chinese domestic context, see: Che Xiuzhen, Jincheng Shang and Jinhu Wang, 'Strategic Environmental Assessment and its Development in China' (2002) 22 *Environmental Impact Assessment Review* 101–109; Cun-kuan Bao, Lu Yong-sen and Shang Jin-cheng, 'Framework and Operational Procedure for Implementing Strategic Environmental Assessment in China' (2004) 24 *Environmental Impact Assessment Review* 27–46; Tao Tang, Tan Zhu and He Xu, 'Integrating Environment into Land Use Planning through Strategic Environmental Assessment in China: Towards Legal Frameworks and Operational Procedures' (2007) 27 *Environmental Impact Assessment Review* 243–265; Olivia Bina, 'Context and Systems: Thinking More Broadly about Effectiveness in Strategic Environmental Assessment in China' (2008) 42 *Environmental Management* 717–733; and Tan Zhu and Kin-Che Lam (eds), *Environmental Impact Assessment in China* (Nankai University and Chinese University of Hong Kong, 2009).

environmental information and public participation for example,[107] this is generally required by Article 2(4), although this merely refers to involvement 'at appropriate stages of the development and environmental planning processes'. Standing requirements of domestic law (cited in Article 3(2)) may inhibit this, including national interpretations of who 'affected citizens' are. Article 1(6) makes further reference to the effect of the SEA and EIA in decision-making; Contracting Parties must coordinate and cooperate with each other so that results 'are taken into account in development planning activities for the Region, and to implement the regional and project-specific EMMPs'.

Article 1(7) states that: 'All EAs and EMMPs will be conducted in accordance with internationally accepted procedures and guidelines'. It is argued that this includes Principle 10 of the Rio Declaration 1992,[108] and that the original soft law nature of this may now have become part of customary international law, especially the requirements for access to environmental information and public participation. Article 1(7) is supplemented by Article 2(1), which states: 'The Contracting Parties will strive, both individually and in cooperation with each other, to meet the objectives of international environmental agreements and norms with respect to their activities in the Region'. There is also an obligation in Article 2(2) to coordinate activities with respect to implementation of national laws and to consider 'proposals for progressive harmonization of such laws', which is designed to establish minimum standards for environmental planning and management. Article 2(3) emphasizes the importance of capacity building for sustainable development, with an obligation to cooperate through personnel exchanges, and technology transfer. Article 2(5) commits the Contracting Parties to provide or seek funding for the preparation of EIA and EMMP documentation and for the other environmental responsibilities.

---

[107] The relationship between public participation and EIA is also addressed in the literature in a Chinese context. See: Jesse L Moorman and Ge Zhang, 'Promoting and Strengthening Public Participation in China's Environmental Impact Assessment Process: Comparing China's EIA Law and U.S. NEPA' (2007) 8 *Vermont Journal of Environmental Law* 281–335; and Shui-Yan Tang, Ching-Ping Tang and Carlos Wing-Hung Lo, 'Public Participation and Environmental Impact Assessment in Mainland China and Taiwan: Political Foundations of Environmental Management' (2005) 41 *Journal of Development Studies* 1–32.

[108] Declaration on Environment and Development (Rio de Janeiro, 14 June 1992) 31 ILM (1992), 874. See Jonas Ebbesson, 'Principle 10: Public Participation', in Jorge E. Viñuales (ed), *The Rio Declaration on Environment and Development: A Commentary* (Oxford University Press, Oxford, 2015).

As indicated, since the establishment of the *Tumen Programme* there have been 16 meetings of the Consultative Commission. At the Eighth Meeting held in Changchun, China in 2005, as well as extending the original 1995 Agreements, governments agreed on a Strategic Action Plan 2006–2015. This focused on GTI activities in four priority sectors: transport, energy, tourism and investment, with environment as a cross-cutting theme. The Ninth Meeting in Vladivostok, Russia in 2007 reaffirmed the commitment to sustainable development in the region, with a special session devoted to joint GTI projects to be implemented. Further institutional structures were agreed on at the meeting, including the CFE. The Tenth meeting was held in Ulaanbaatar, Mongolia in 2009,[109] although there was no representative from North Korea present and North Korea's involvement effectively ceased thereafter.[110]

Within the framework of the Strategic Action Plan, the Secretariat has prepared biennial work plans for submission to the Commission. These have focused on implementing projects of practical value to member governments, with the importance of private sector funding emphasized. Prioritization has been given to transport infrastructure, facilitating movement of goods through the seaports in the region. Studies have been conducted on the sea-land routes of northeast Asia,[111] and the transport corridors in the subregion,[112] the latter with a focus on financing infrastructure development. The environment does however receive particular attention, and in 2004 it had attracted 85 per cent of total programme funding. A comprehensive Strategic Action Plan (also known as the 'TumenNET') established policy priorities and national action commitments, especially in the areas of biodiversity loss and water issues.[113]

---

[109] Note that Mongolia has its own issues in relation to development of the high country in its territory, see: Andrew Laurie, Jargal Jamsranjav, Onno van den Heuvel and Erdenesaikhan Nyamjav, 'Biodiversity Conservation and the Ecological Limits to Development Options in the Mongolian Altai: Formation of a Strategy and Discussion of Priorities' (2010) 29(3) *Central Asian Survey* 321.

[110] Current membership therefore excludes North Korea. See: <http://www.tumenprogramme.org/?list-1528.html> accessed 22/6/16.

[111] Sung-woo Lee, Geun-sub Kim and Eun-woo Kim, *Evaluation Study on the Sea-Land Routes in Northeast Asia* (Greater Tumen Initiative/Korea Maritime Institute, Beijing, 2014).

[112] Marcel van den Broek, *Transport Corridors in the Greater Tumen Region: Financing Infrastructure Development: Study Report* (Greater Tumen Initiative/German Cooperation, 2014).

[113] For background and evaluation, see Myungjin Kim, above n. 102, 198–200. See also See-Won Byun, above n. 79.

One result of the Strategic Action Plan, as funded by the Global Environment Facility (GEF) and the UNDP, was a series of EIA strategy workshops, the first held in April 2002, which recommended that an SEA/EIA Training Course standardize the content of the baseline data concerning trans-border projects to meet the minimum requirements of affected countries. Initiated by South Korea, workshops for 'Transboundary Environmental Impact Assessment in Northeast Asia Region' have been held regularly since 2004, with the 2009 workshop held in China. The purpose has been to standardize procedure and enhance cooperation in relation to the projects planned. Supported also by the UNECE, workshops were also held in 2012 and 2013, attended by international participants.[114]

GTI Projects approved at the Ninth Consultative Commission Meeting were discussed during the Tenth Consultative Commission Meeting in Ulaanbaatar in 2009. The two environment projects approved are, first, the GTI Environmental Cooperation, which focuses on transboundary EIA and environmental standardization in northeast Asia,[115] and, second, the Feasibility Study on the Tumen River Water Protection. An earlier Feasibility Study on Establishing the Lower Tumen Area Trans-boundary Biosphere Reserve commenced in 2002.[116] Other projects included the Mongolia-China Railway Feasibility Study, which is a proposal for a

---

114 Korea Environment Institute, Ministry of the Environment Korea, UNECE Espoo Convention, *Workshop on Transboundary Environmental Impact Assessment in East Asia*, Seoul, South Korea, 13–15 June 2012; Korea Environment Institute, Ministry of Environment, UNECE Espoo Convention, *Workshop on Environmental Impact Assessment in a Transboundary Context in Eastern Asia*, May 2013.

115 See Paper 1A, Session 1, Overview of GTI Projects, The 10th Consultative Commission Meeting for the Greater Tumen Initiative (GTI), 24–25 March 2009, Ulaanbaatar, Mongolia, in particular 16–17.

116 This was aimed at developing a cooperative transboundary mechanism for biodiversity conservation, (including endangered tigers and leopards), potentially to result in a Trans-boundary Biosphere Reserve (TBR) in the Lower Tumen River Area for inclusion in the UNESCO Man and Biosphere Trans-boundary Biosphere Reserve list. See Ian Davies and Valery Neronov, Field Diagnostic and Planning Mission to the Lower Tumen River Area of China and the Russian Federation, May 26–June 7 (2003), 84–95 for background; see also United Nations Development Program and Korean National Commission for UNESCO, *Project Final Report, Feasibility Study on the Establishment of the Lower Tumen River Area Transboundary Biosphere Reserve* (2004). For academic comment see Sangmin Nam, 'Ecosystem Governance in a Cross-border Area: Building a Tumen River Transboundary Biosphere Reserve' (2005) 7 *China Environment Series* 83–88.

railway line linking Mongolia with the seaports in the region, and the Road and Harbour Project on the China-DPRK border, which includes the reconstruction of a trans-border road into a new road and bridge.

The environment was not mentioned at the Eleventh Consultative Commission Meeting, held in Changchun, China in 2010. This was also the case regarding the Twelfth Consultative Commission Meeting, held in Pyeongchang, South Korea, where there was only one mention of 'sustainable development' and all three 'priority sectors' (transport, trade facilitation and tourism) noted with no comment on the 'cross-cutting' environment issues highlighted at the Eighth Meeting. At the Thirteenth Meeting in Vladivostok in 2012, two sentences were devoted to the cross-cutting issues of energy and the environment, where the member governments stated:

> We reaffirmed the importance of regional cooperation on energy and environment, and recognized the need to develop and conduct more practical and result-oriented cooperation projects with the help of the GTI Secretariat. We further tasked the GTI Secretariat to facilitate the advancement of the existing projects agreed upon multilaterally and bilaterally within the GTR region.[117]

Whether this provided the initiative needed to reinvigorate environmental issues in the *Tumen Programme* is, however, unlikely. At the Fourteenth Meeting, held in Ulaanbaatar in 2013, it appeared to be 'business as usual', with only reference made to activity under the Energy Board and advocacy to the effect that: 'We expect more active engagement of relevant energy administrations from GTI member countries in the cooperation'.[118] In relation to the environment however, there was only silence, despite activity in the transport, trade and investment and tourism priority sectors with potentially significant negative environmental effects.

The Fifteenth Meeting was held in Yanbian, China in 2014, and four proposals for energy[119] projects were approved: a power generation and

---

117 Vladivostok Declaration, 10 October 2012, 2.

118 Ulaanbaatar Declaration, 30 October 2013, 2.

119 In relation to energy, note Ryu Ji Chul, *GTI Energy Cooperation Programme in Northeast Asia* (Greater Tumen Initiative, 2016). This does not address environmental issues other than to comment for example that: 'Recently, environment became an important factor in shaping the energy mix structure in the countries in NEA. The importance of less carbon intensive fuel or carbon-free energy sources, namely, natural gas and renewable energy sources, is well-recognized in order to minimize the environmental effect caused by the fossil fuel use' (at 21).

transmission project (a preliminary economic analysis of power generation and transmission) through the Korean peninsula; a feasibility study of cooperation on adjusting from coal to synthetic natural gas; an industrial energy efficiency workshop; and research on regional industrial energy efficiency cooperation. The environment also received much greater attention, with the comment from the member countries – albeit in the absence of concrete action to 'promote environmental sustainability in the GTR while conducting economic activities'[120] – as follows:

> We attach great importance to enhance the GTI's cooperation in the green growth and environment protection. We agree to carry on the cooperation activities in the fields of green technology, public awareness on environment protection, enterprises' social responsibilities, green investment and green finance, as well as bio diversity, and to select flag GTI environment projects to consolidate support from various environment institutions in NEA [Northeast Asia].[121]

3.2.2 Air pollution arrangements

Chapter 4 addressed air pollution in the southeast Asian context with reference to the *Haze Pollution Agreement*.[122] While the countries of south Asia and east Asia have yet to produce a comparable binding agreement, there have been policy developments in recognition of the significance of the problem in other parts of the Asian region. As indicated in Chapter 5 above, south Asian states produced the *Malé Declaration* to highlight concerns about air pollution. In east Asia, there is also awareness of the problem in other countries that have suffered from the effects of transboundary air pollution,[123] notably among policy-makers, scientists and citizens in some parts of northeast Asia, primarily Japan, South Korea and China.

These three countries host Tripartite Environment Ministers Meetings on a regular basis to progress developments in this and other areas of common concern.[124] An initial Action Plan (2010–2014) was released in

---

[120] As provided for by the Strategic Action Plan, para 4 'Strategic Objectives', 2.

[121] Yanbian Declaration, 17 September 2014, 3.

[122] Agreement on Transboundary Haze Pollution (Kuala Lumpur, 10 June 2002), in force 25 November 2003.

[123] See for example in relation to Japan and Korea, Inkyoung Kim, 'Environmental Cooperation of Northeast Asia: Transboundary Air Pollution' (2007) 7(3) *International Relations of the Asia-Pacific* 439–462.

[124] For a Chinese report of the most recent meeting – which did not however address air pollution issues – held in Shizuoka, Japan on April 27 2016 see:

Hokkaido in May 2010, but air pollution specifically was not one of the ten priority areas of environmental cooperation in this plan; climate change and pollution control generally were however included as part of the overall aspirations and efforts. In the second Action Plan (2015–2019) air pollution is included as the first of the priority areas.[125] However, despite these efforts, in this subregion there has been a significant lack of consensus on the environmental impacts of transboundary air pollution at both domestic and subregional levels. Juli Kim comments for example: 'the Korean Peninsula and Japan have felt the brunt of China's "export" through acid rain, mercury, and other airborne contaminants as Siberian winds and dust storms flush out China's pollution every spring, severely damaging forests and watersheds'.[126]

Recent reports note the adverse effects of this pollution upon North America, combined with the weather patterns over the Pacific Ocean,[127] and that this has been generated by high levels of particulate pollution in China, India and Japan. As a result (non-binding) agreements on air pollution have been produced to counter the trend, including the *Tripartite Joint Action Plan on Environmental Cooperation between Korea, China and Japan* (2010).[128] Policy and science-related initiatives, such as

---

<http://english.mep.gov.cn/About_SEPA/leaders_of_mep/chenjining/Activities/201605/t20160511_337500.shtml> accessed 24/6/16. For a summary of the discussions held at previous meetings, see: <https://www.env.go.jp/earth/coop/coop/english/dialogue/temm.html> accessed 24/6/16.

125 Tripartite Environment Ministers, Tripartite Joint Action Plan on Environmental Cooperation (2015–2019), April 2015.

126 Juli S Kim, 'Transboundary Air Pollution – Will China Choke on its Success?' (2007) *China Environmental Health Project Fact Sheet*, 2. See also Inkyoung Kim, above n. 123, 439.

127 For example AFP/Jili, 'Air Pollution in Asia Intensifies Cyclones', *The Japan Times*, 1 February 2014; William Wan, 'Study: Pollution from Chinese Factories is Harming Air Quality on US West Coast', *The Washington Post*, 21 January 2014. For the recent situation, see also Inkyoung Kim, 'Messages from a Middle Power: Participation by the Republic of Korea in Regional Environmental Cooperation on Transboundary Air Pollution Issues' (2014) 14 *International Environmental Agreements* 147–162; and Jae-Hyup Lee, 'Transboundary Pollution in Northeast Asia: An International Environmental Law Perspective' (2013) 35 *University of Hawaii Law Review* 769–785.

128 Between Japan, South Korea and China. For the latest developments, see <https://www.env.go.jp/press/file_view.php?serial=22137&hou_id=16621> accessed 19/11/14.

the East Asia Network on Acid Deposition (EANET),[129] the Joint Forum on Atmospheric Environment in Asia and the Pacific,[130] the Regional Forum on Environment and Health in Southeast and East Asian Countries (Thematic Working Group on Air Quality) and Clean Air Asia, have also helped raise awareness through scientific monitoring, information exchange, conferences and capacity-building activities.

Another initiative – which applies both to east and south Asia – is the Greenhouse Gas and Air Pollution Interactions and Synergies (GAINS-Asia) scientific model, which was developed by experts in China and India in association with the International Institute for Applied Systems Analysis. This identifies the local health impacts of various transboundary air pollutants, and analyses the cost-effectiveness of potential tools for reducing them.[131] The GAINS-Asia model is being applied extensively in China, India and Pakistan, and is freely accessible online for use by other countries.[132] Key policy messages from GAINS-Asia include that:

- Economic growth will exacerbate the current air quality problems in Asia unless existing pollution control laws are significantly upgraded.
- Compared to a conventional strategy, the GAINS cost-effectiveness approach can find ways to reduce costs for improving air quality by 80%.
- By selecting a mix of measures to simultaneously cut air pollution and greenhouse gas emissions, countries can also further reduce air pollution control costs.[133]

---

[129] Participating countries are China, Vietnam, Cambodia, Laos, Thailand, Indonesia, Malaysia, Philippines, Japan, Mongolia, South Korea and Russia.

[130] This initiative began in 2010, with the release of the *Joint Plan for Joint Activities on Air Pollution in Asia and the Pacific*.

[131] One key finding to date has been that the application of advanced emission control technologies could reduce health impacts in China by 43 per cent by the year 2030.

[132] Richard Ballaman, 'Comparing Regional Environmental Governance in East Asia and Europe – Air Pollution', in Jörg Balsiger and Aysun Uyar (eds), *Comparing Regional Environmental Governance in East Asia and Europe: Proceedings* (Research Institute for Humanity and Nature, Kyoto, 2013) 71–74; see also in the same volume Esook Yoon, 'Governing Transboundary Pollution in Northeast Asia' 65–70; and Sangmin Nam and Heejoo Lee, 'Reverberating beyond the Region in Addressing Air Pollution in Northeast Asia' 75–89.

[133] See: 'GAINS-Asia' <http://gains.iiasa.ac.at/gains/ASN/index.login?logout=1> accessed 22/11/14.

The North-East Asian Subregional Programme for Environmental Cooperation (NEASPEC)[134] also has a role to play in addressing transboundary air pollution in northeast Asia.[135] As a part of the UN Economic and Social Commission for the Asia Pacific (UNESCAP), like the UNECE and its *Air Pollution Convention*,[136] it is also not a focus of this book.[137] A NEASPEC project currently being undertaken by Russia aims to study existing subregional and international mechanisms for transboundary air pollution and to identify gaps in knowledge, prioritize needs and propose options for a strengthened multilateral approach.[138] According to Ballaman, it was 'expected that the developments in Northeast Asia might lead to the establishment of a policy-type umbrella framework that would cover the existing monitoring and modeling programs like the Joint Research Project on Long-Range Transboundary Air Pollutants and the Acid Deposition Monitoring Network in East Asia'.[139]

3.2.3 *East Asian Regional Seas Programme*

Of the four Regional Seas Programmes analysed in this book, this is the only one directly administered by UNEP;[140] the *South Asian Seas Programme* considered earlier in this Chapter is administered by SACEP. Of the others, the *Red Sea/Gulf of Aden* regime is administered by PERSGA, and the *Arabian Gulf/Gulf of Oman* regime by ROPME, both of which were considered in Chapter 2; while the *Caspian Sea* regime is an independent programme that – unlike the others – was not established under the auspices of UNEP. The *East Asian Seas Programme* is also not specific to East Asia as defined by this chapter, but includes states in southeast Asia and Australia; as such, it is also relevant to Chapter 4. The east Asian states are China and South Korea.

This varied range of states is matched by an equally diverse environment of crowded straits, many islands, wide gulfs, shallow estuaries and

---

134 <http://www.neaspec.org/?ckattempt=1> accessed 24/6/16.

135 See for example, United Nations Economic and Social Commission for Asia and the Pacific, *NEASPEC Workshop on Transboundary Air Pollution in North-East Asia, Report of the Meeting*, 10–11 November 2011, Incheon.

136 Convention on Long-Range Transboundary Air Pollution (Geneva, 13 November 1979) 1302 UNTS 217, in force 16 March 1983.

137 In contrast, see Marsden and Brandon, above n. 3, where this is the focus.

138 Marsden and Brandon, above n. 3. See also: <http://www.neaspec.org/our-work/transboundary-air-pollution-mitigation> accessed 24/6/16.

139 Above n. 132, 73.

140 <http://www.unep.org/regionalseas/programmes/unpro/eastasian/default.asp> accessed 19/3/15.

a large number of heavily populated countries, where fish is an essential part of the diet and fishing forms a key part of people's livelihoods. There are also numerous threats including erosion and siltation from land development, logging and mining, blast fishing in coral reefs, conversion of mangroves, overfishing, unimpeded coastal development and disposal of untreated wastes. Seven areas of focus were identified for the region:

- Developing and maintaining a regional database (later changed to a regional metadata base);
- Promoting, improving, networking and maintaining marine protected areas in the region;
- Implementing activities to restore marine habitats;
- Assisting with State of Environment reporting for agencies preparing such reports and undertaking marine and coastal assessment;
- Implementing activities to reduce land-based sources of pollution;
- Encouraging monitoring and environmental assessment, including mapping in the region; and
- Encouraging and implementing projects to build capacity in the member countries to counter environmental degradation and to educate all members of the community in caring for the marine resources of the region.[141]

The Action Plan for the Protection and Development of the Marine and Coastal Areas of the East Asian Region was approved in 1981; initially subregional, it involved five ASEAN states with five more welcomed in 1994. Although the programme is UNEP administered, the Action Plan is coordinated from Bangkok by the Coordinating Body on the Seas of East Asia (COBSEA).[142] In the absence of a treaty, the programme promotes compliance with existing environmental agreements based on goodwill.

Long-term strategies (1987–96) for the East Asian Seas Action Plan were adopted in 1987. A subsequent long-term strategy of COBSEA (1994–2009) was adopted in 1994; supplemented by the document, 'Vision and Plan – A Systematic Approach' adopted in 1999.[143] Following the tsunami experienced by this subregion in 2004, the urgent need for an early warning system and improved disaster management to reduce the loss of human lives and property was recognized. The long-term

---

141 Ibid.

142 All relevant documents are accessible at the following site: <http://www.cobsea.org/publications.html> accessed 24/6/16.

143 Ibid.

future depends however on the rehabilitation of damaged marine ecosystems, and the avoidance and management of future oil pollution.[144] The success of all of these things is dependent upon the compliance of the member states with their obligations in international law. China's position in relation to the South China Sea, which was raised in Chapter 4, does not provide any confidence in its abilities in this respect, given the damming findings of the Permanent Court of Arbitration (PCA) under the LOSC in July 2016:

> Based on the compelling evidence, expert reports, and critical assessment of Chinese claims … the Tribunal has no doubt that China's artificial island-building activities on the seven reefs in the Spratly Islands have caused devastating and long-lasting damage to the marine environment. The Tribunal accordingly finds that through its construction activities, China has breached its obligation under Article 192 to protect and preserve the marine environment, has conducted dredging in such a way as to pollute the marine environment with sediment in breach of Article 194(1), and has violated its duty under Article 194(5) to take measures necessary to protect and preserve rare or fragile ecosystems as well as the habitat of depleted, threatened or endangered species and other forms of marine life.[145]

4. THE THIRD POLE

4.1 Introduction

It is clear from an examination of the geographical spread of the subregional environmental regimes in Asia that despite a number of relevant agreements in place, the subregion currently without a detailed environmental governance regime is south Asia; similarly, while in east Asia the *Tumen Programme* may contain elements furthering environmental protection, this is but one aspect of this regime, which is overwhelmingly focused on development. This section will briefly evaluate the potential for the establishment of a transboundary environmental regime in the mountainous area known as the 'Third Pole', which is

---

[144] See Robin Warner, 'Stemming the Black Tide: Cooperation on Oil Pollution Preparedness and Response in the South China Sea and East Asian Seas' (2015) 18(2) *Journal of International Wildlife Law* 184–197.

[145] PCA, above n. 84, para 983.

constituted mainly by the Himalayas and Tibetan Plateau[146] and comprises numerous transboundary watercourses; it will furthermore make recommendations for the development of such a regime, with suggestions for future research outlined at the end of this chapter, which also concludes the book.

In summary, the analysis of the existing regimes in southwest, central, southeast and northeast Asia indicates that a focus on developing and establishing effective regulatory frameworks for biodiversity conservation and transboundary EIA are likely to gain the most support, as they are a common part of the regimes outlined in Chapters 2–4, and to some extent the *Tumen Programme* in this chapter also. This is enhanced in the Third Pole subregion by the multilateral arrangements for biodiversity conservation, which as they operate in a cross-border context,[147] therefore contend with the transboundary dimensions required to some extent currently. Similarly, in relation to transboundary EIA, while this may not be a part of the current measures between the Third Pole states, it is however a requirement of general international law, and is often required by the MDBs/IFIs.[148]

---

[146] Other mountain ranges forming part of this subregion include the Pamirs and Karakoram. See: Thomas Breu, Hans Hurni and Andrea Wurth Stucki, *The Tajik Pamirs: Challenges of Sustainable Development in an Isolated Mountain Region* (Centre for Development and Environment, University of Berne, Berne, 2003); Michelle Lim, 'Laws, Institutions and Transboundary Pasture Management in the High Pamir and Pamir-Alai Mountain Ecosystem of Central Asia' (2012) 8(3) *Law, Environment and Development Journal* 43; K Hewitt, *Glaciers of the Karakoram Himalaya: Glacial Environments, Processes, Hazards and Resources* (Springer, 2014); and Interstate Sustainable Development Commission/Regional Mountain Centre for Central Asia, *Draft Strategy of the Regional Mountain Centre of Central Asia* (Bishkek, 2009).

[147] See Convention on Biological Diversity, above n. 5, and, for comment in the Third Pole context, Simon Marsden, 'Asian Precedents for Effective Biodiversity Conservation in the Third Pole', paper presented to *International Symposium: Biodiversity, Sustainable Development and the Law*, Centre for International Sustainable Development Law/Lauterpacht Centre for International Law, St John's College, University of Cambridge, 2015.

[148] See generally above n. 86 and David B Hunter, *International Financial Institutions and International Law* (Kluwer Law International, 2010); in the context of central Asia, Mehrdad Nazari, 'The Transboundary EIA Convention in the Context of Private Sector Operations Co-financed by an International Financial Institution: Two Case Studies from Azerbaijan and Turkmenistan' (2003) 23 *Environmental Impact Assessment Review* 441, and for more detail than this section provides in connection with the Third Pole, Simon Marsden, 'Environmental Assessment of Cross-Border Development: China and the Third

Importantly, the domestic law of China, India, Pakistan and the other states in the subregion has no reported provision for the evaluation of significant environmental effects upon their neighbours; bilateral agreements, in particular the *Indus Waters Treaty* between India and Pakistan,[149] as seen, also fail to deal with such matters adequately or at all. Regional agreements – such as the UNECE *EIA Convention*[150] and *Water Convention*[151] – also apply barely or not at all in the Third Pole subregion.[152] This may however change, since both of these agreements are now open to a global membership which includes these states.[153] The next section will define the Third Pole in detail, together with the political constraints to subregional governance, while the sections that follow outline the main concerns in relation to water management and biodiversity protection, and infrastructure development respectively. As will be seen, both of these areas are informed by – and involve the application of – transboundary EIA as part of the regulatory approach.

4.2 Defining the Third Pole and Political Constraints to Subregional Governance

The Third Pole is primarily constituted by the Himalayas and Tibetan Plateau. However, it also includes the Hindu Kush Mountains, from central Asia to Kashmir at the Pakistan/India border; from the northwest, the Pamir of Tajikistan; from the west, the Karakoram of Afghanistan and Pakistan; and extends through the Himalayas to the east, where the Kunlun Mountains also border the Tibetan Plateau.[154] This range of

---

Pole' (2016) 18 *Journal of Environmental Assessment Policy and Management* 1650009-1-1650009-21.

149 See Sarfraz, above n. 50, 204–216; and Wolf and Newton, above n. 48.

150 Above n. 86.

151 Above n. 53.

152 Moynihan and Magsig, above n. 55. Water security is however a major issue here as in the other Asian subregions; see Bjørn Oliver Magsig, *International Water Law and the Quest for Common Security* (Earthscan/Routledge, Abingdon, 2015) 144–199, which focuses upon Himalayan Asia; and Morton above n. 64.

153 See UNECE, above n. 86 (Espoo), and United Nations Economic Commission for Europe, Convention on the Protection and Use of Transboundary Watercourses and Lakes, *The Global Opening of the 1992 UNECE Water Convention* (UNECE, 2013).

154 N Chettri, AB Shrestha, Y Zhaoli, B Bajracharya, E Sharma and H Ouyang, '"Real World" Protection for the Third Pole and its People', in Falk Huettmann (ed), *Protection of the Three Poles* (Springer, 2012) 113.

geographical terrain is illustrative of the challenges of ensuring sustainable development across a very broad subregion.[155] It is high altitude, as the 'First' and 'Second' Poles (the Arctic and Antarctic) are high latitude, and is similarly affected by climate change; for example there has been a loss of a fifth of the glaciers in China since the 1950s, with major effects upon water supply as a result.[156] It is also similar to the Arctic in the focus on sustainable development and the national development pressures in both of these 'Poles'; it is fundamentally unlike the Antarctic however, in that there, firstly, attention is focused upon environmental protection, and, secondly, and unlike the Arctic also, the Antarctic is not the sovereign territory of nation states.[157]

Chettri et al. point to three key points of difference between the Third Pole and the First and Second Poles: significant population, climatic differences and rich biodiversity;[158] the biodiversity alone makes it worthy of particular attention, and explains why governance in this area is so important. Another difference worth noting is the earthquake-prone nature of the subregion, with major incidents in Nepal in 2014, and Afghanistan and Pakistan in 2015.[159] New scientific developments may

---

155 Breu et al., above n. 146. Two organizations have played an important role in highlighting the issues faced by the Third Pole: the International Centre for Integrated Mountain Development (ICIMOD) and the Third Pole Environment (TPE). ICIMOD is based in Kathmandu and is a regional intergovernmental centre which was established over thirty years ago by Afghanistan, Bangladesh, Bhutan, China, India, Myanmar, Nepal and Pakistan as a means of cooperating in relation to the sustainable development of the Hindu-Kush-Himalaya subregion. TPE, based in Beijing, is a means of information exchange holding annual scientific workshops across the states of the subregion and globally, with the most recent, the sixth, held in Columbus, USA, and which opened an office in the USA emphasizing its global significance. See: David Molden and Eklabya Sharma, 'ICIMOD's Strategy for Delivering High-quality Research and Achieving Impact for Sustainable Mountain Development' (2013) 33(2) *Mountain Research and Development* 179–183.

156 Jing Li, 'A Fifth of Glaciers Lost Since the 1950s', *South China Morning Post*, 15 December 2014.

157 In relation to the Arctic, see generally Michael Byers, *International Law and the Arctic* (Cambridge University Press, Cambridge, 2015); and to Antarctica, Ben Saul and Tim Stephens (eds), *Antarctica in International Law* (Hart Publishing, Oxford, 2015).

158 Above n. 154, 114–115.

159 Soutik Biswas, 'Nepal Earthquake: How India and China Vie For Influence', *BBC News*, 28 April 2014; Boone, above n. 36.

help with predictions in the future;[160] however, it is also now recognized that while most earthquakes are triggered by natural causes, man-made earthquakes ('induced seismicity') due to mining, dam-building and fracking have also resulted, and are becoming more common.[161] This is a major concern given the pro-development focus of China in the subregion, and is supported by scientific claims for example that dam-building led to the Sichuan earthquake in 2008.[162]

The need to establish and improve collaborative governance and plan for such events has attracted the attention of only a few scholars, in an area with a significant population density of towns and cities. Contending with the added challenges of climate change,[163] water management (including, issues linked with climate change, flooding)[164] and Indigenous/minority rights adds to the significance of the issues to be faced. Examples of the future pressures that this subregion is likely to be subject to are the extraction of oil and gas reserves, and further development of the transboundary rivers originating in the Tibetan Plateau for hydro-electricity. The challenges seen in relation to hydro-electric development in central and southeast Asia are illustrative of what is likely in this respect. It is noted that India, Pakistan, Nepal and Bhutan collectively have plans for around 400 dams, and China 100; within India 160 dams are proposed for Assam specifically.[165]

Politically, the Tibetan Plateau is part of the 'Tibet Autonomous Region', in turn part of China. As China has assumed a leadership role in relation to development at home and abroad,[166] it is to China that states in south Asia increasingly look. However, whether it is well placed to sustainably manage these watercourses and the other resources of the subregion is in doubt because of its development focus. As with all states,

---

160 Stephen Chen, 'Skin-Deep Explanation for Tibet's Bulging Plateau', *South China Morning Post*, 24 January 2015; Kees Vuik, 'Is it Finally Possible to Predict Earthquakes?', *The Guardian*, 20 May 2015.

161 Lovett, above n. 59.

162 Moore, above n. 59.

163 See Morton, above n. 64. See also John Vidal, 'Most Glaciers in Mount Everest Area will Disappear with Climate Change – Study', *The Guardian*, 27 May 2015.

164 Mirza, above n. 18.

165 Kieran Cooke, 'The Dams of India: Boon or Bane?', *The Guardian*, 17 March 2014; John Vidal and K Dasgupta, 'Himalayas at Risk in Hydro Dam Water Grab', *New Zealand Herald News*, 13 August 2013.

166 Sonny Lo, 'Yesterday's Middle Kingdom, China Today Seeks to be the Heartland of Asia', *South China Morning Post*, 8 December 2014.

China must however comply with international law in terms of development, including transboundary EIA.[167] In relation to these other resources, the search for minerals in the Himalayas has begun,[168] and this must be managed very carefully if water pollution is not to result. It is however air pollution that casts the largest current shadow over the Third Pole,[169] with glacier and snow melt arguably of greater significance here than in the Arctic or Antarctic because of the direct, immediate effect upon the large populations that rely on the rapidly reducing waters, or because of the unpredictability of environmental flows that increasingly result in major flooding incidents.

The Third Pole is therefore found in the northern part of south Asia where it connects with east Asia because of the link between the mountains and the Tibetan Plateau. Because of this link, although it is referred to throughout as a subregion, technically it is a 'cross'-subregion as it links the south and east Asian subregions. Unlike the Arctic, Caspian Sea and Red Sea/Gulf of Aden, which are cross-regions, it is squarely confined to the Asian continent. It is however in urgent need of policy and legal responses to avoid or mitigate the pressures outlined above and below; yet to date, these responses have been limited by the geopolitics of the primary players in the subregion. This has seen China, India and Pakistan in dispute over the frontiers of the Himalayas for example. In the case of India and Pakistan, this conflict has a long history, since independence from former UK colonial rule in the 1940s;[170] in the case of India and China, border disputes began in the 1960s, and continue to cloud relations.[171]

---

[167] See above nn. 86 and 87.

[168] Shishir Prashant, 'Jammu-Pak Border Might Have Oil and Gas, Says ONGC', *Business Standard*, 7 February 2014.

[169] See Lu et al., above n. 11.

[170] Aditya Kalra and Asad Hashim, 'India-Pakistan Peace Talks Collapse, Deadlock Sours Relations', *Reuters*, 23 August 2015.

[171] Anthony Carty and Fozia Nazir Lone, 'Some New Haven International Law Reflections on China, India and their Various Territorial Disputes' (2011) 19(1) *Asia Pacific Law Review* 93; for recent commentary, see J Burke, 'India-China Border Standoff Highlights Tensions Before Xi Visit', *The Guardian*, 16 September 2014; Mini Chan, 'Agreements Reached on a Number of Issues, But No Progress on Long-running Border Row', *South China Morning Post*, 21 September 2014.

4.3 Water Management and Biodiversity Protection

The cross-border effects of development proposals have been highlighted throughout this book as of major significance for subregional environmental governance, in whichever of the subregions they are found; they are however of particular concern in the Third Pole subregion because of the absence of any transboundary environmental regime.[172] In connection with water management, there is an extensive literature on this,[173] which has also given broad coverage to transboundary EIA.[174] As such, and because of the discussion in previous chapters, this section will be fairly brief, concentrating on those areas relevant to the Third Pole; infrastructure development issues will also be considered further in the next section.

Proposals for new development in this subregion include major hydroelectric dams of watercourses originating in the Tibetan Plateau such as the Ganges, Indus and Brahmaputra; these are often referred to as 'Water Towers' for their location and significance.[175] The origin of the Mekong River is also in the Tibetan Plateau, and the inadequate regulation of dams in southeast Asia[176] that was seen in Chapter 4, emphasizes the need for improved regulation of development proposals for all the south Asian rivers, to avoid the predominance of national economic interests

---

[172] See Marsden, above n. 148.

[173] In addition to references elsewhere in this book, see also: Ariel Dinar, Shlomi Dinar, Stephen McCaffrey and Daene Mckinney, *Bridges over Waters: Understanding Transboundary Water Conflict, Negotiation and Cooperation* (World Scientific, Singapore, 2013); as well as analysing the situation in the Mekong and the Aral Sea, this also discusses the relationship between India and Bangladesh in the Ganges Basin. See also: JR Daniel, SL Pinel and J Brooks, 'Overcoming Barriers to Transboundary Water Governance' (2013) 33(3) *Mountain Research and Development* 215–224.

[174] See for example: Carl Bruch, Mikiyasu Nakayama, Jessica Troell, Lisa Goldman and Elizabeth Maruma Mremaet, 'Assessing the Assessments: Improving Methodologies for Impact Assessment in Transboundary Watercourses' (2008) 26 *Impact Assessment and Project Appraisal* 241–242; and Angela Z Cassar and Carl E Bruch, 'Transboundary Environmental Impact Assessment in International Watercourse Management' (2003) 12 *New York University Environmental Law Journal* 194–195.

[175] Chettri et al., above n. 154; and Wouters, above n. 57, 71.

[176] Save the Mekong Coalition, 'Request for clarification regarding the First Regional Consultation on the Don Sahong Hydropower Project', Letter to Mr. Hans Guttman CEO of the Mekong River Commission, 11 November 2014.

that has subsumed subregional environmental protection.[177] Arguably, this should include completion of detailed SEAs of the policies, plans and programmes for the river basins concerned, and compliance with any recommendations made.[178]

Another example of the challenges to be met in connection with both water management and biodiversity protection can be seen in regard to international cooperation in the Brahmaputra basin. Since the Brahmaputra River carries more water than the 20 largest rivers in Europe combined and is only exceeded in volume by the Amazon and the Congo in south America and Africa, the electricity generating potential of this river has led to 'hydropower experts see[ing] it as an Eldorado of energy potential'.[179] However, unstable geology and unsustainable mining practices on the Tibetan Plateau are resulting in ecological degradation on a massive scale.[180] The governance situation has been summarized by Hilton in the following terms: 'The lack of a cooperative framework for managing the whole river systems has left it vulnerable to short term, competitive exploitation as rival neighbours race to stake pre-emptive claims'.[181]

Similar challenges can also be seen with respect to the other transboundary watercourses analysed in this chapter – the Indus – but also the Ganges, which shares the basin with the Brahmaputra. There is furthermore a close connection between water management and biodiversity protection.[182] This has been considered by scholars in regard to central Asia,[183] the Mongolian

---

[177] Tun Myint, *Governing International Rivers: Polycentric Politics in the Mekong and the Rhine* (Edward Elgar Cheltenham, 2012).

[178] In relation to the Mekong, see International Center for Environmental Management, *MRC Strategic Environmental Assessment of Hydropower on the Mekong Mainstream: Final Report* (Mekong River Commission, Hanoi, 2010). Note that at the time of finalization of the manuscript, the MRC published the *Strategic Plan 2016–2020*, which 'identifies basin-wide development and management needs, challenges, opportunities and priorities'. See: <http://www.mrcmekong.org/highlights/strategic-plan-2016-2020/> accessed 22/7/16.

[179] Hilton, above n. 42, 4.

[180] Hilton, above n. 42, 5.

[181] Hilton, above n. 42, 6.

[182] See Marsden, above n. 147.

[183] Michelle Lim, 'Effective Governance across Central Asian Boundaries: What Lessons does the Central Asian Transboundary Watercourse Management Experience Provide for Mountain Biodiversity Conservation across the Kyrgyz-Tajik Boundary?', in Robin Warner and Simon Marsden (eds), *Transboundary Environmental Governance: Inland, Coastal and Marine Perspectives* (Ashgate Publishing, Farnham, UK, 2012) 111.

Altai of east Asia,[184] and in southeast and south Asian contexts.[185] The next section – and the final one before the conclusions – considers in some detail the role of international lending institutions in supporting infrastructure development potentially impacting upon transboundary watercourses and biodiversity. In accordance with the focus of this book and the sections at the end of Chapters 2–4, the role of China is particularly important and is dealt with at length.

4.4 Infrastructure Development and the Role of International Lenders

Examples of infrastructure development have been considered in detail in this book and chapter, with hydro-electric dams for energy production a particular concern for the environmental protection of the Third Pole, as noted.[186] Transboundary EIA is a highly relevant tool for managing environmental effects of a cross-border nature; although it does not currently apply to this subregion, the *Espoo Convention* is the best known of the regional and global agreements for this, and it has spurred developments in other Asian subregions, as seen. The principles and obligations it contains should therefore supplement those found in existing treaties, for example the *Indus Waters Treaty*, as outlined earlier in this chapter.

The principle of transboundary EIA is also now customary international law[187] and must therefore be adhered to in the absence of any treaty provision. The *Espoo Convention* furthermore provides a valuable precedent for the environmental governance options which can evaluate significant environmental effects of development in the Third Pole. This has been the case in regard to most of the other Asian subregions, notably for proposals affecting the watercourses in central Asia (the Caspian Sea), southeast Asia (the Mekong) and northeast Asia (the Tumen). Without applying its procedures effectively, China (and India) for example may risk international claims when environmental harm results to other states, which a focus on precaution and risk management at an earlier time could potentially avoid.

---

[184] Laurie et al., above n. 109.

[185] Price et al., above n. 22.

[186] For perhaps the most extreme example suggested, see: Haroon Saddique and Jason Burke, 'China May Build Rail Tunnel Under Mount Everest, State Media Reports', *The Guardian*, 9 April 2015.

[187] See *Pulp Mills*, above n. 86, and discussion by the numerous commentators cited at that note.

This was considered in some detail at the beginning of the section on east Asia above, and applies for example in the context of the *Tumen Programme*, as analysed above. In the findings of the PCA in relation to the South China Sea, the following was said to illustrate a distinct lack of compliance with international law and its domestic implementation, notably with respect to the CBD, the Convention on International Trade in Endangered Species (CITES),[188] and LOSC:

> Although a Chinese official stated in May 2015 that, as a State party to the CBD and to CITES, 'China will strictly observe provisions of the conventions and honour her obligations in good faith', The Tribunal has seen no evidence that Chinese fishermen involved in poaching of endangered species have been prosecuted under Chinese law.[189]
>
> The Tribunal finds that China has also breached its obligation to protect and preserve the marine environment in respect of its toleration and protection of the harvesting of giant clams by the propeller chopping method.[190]
>
> [T]he obligation to communicate is, by the terms of Article 205 [LOSC], to 'competent international organizations, which should make them available to all States.' Although China's representatives have assured the States parties to the Convention that its 'construction activities followed a high standard of environmental protection,' it has delivered no assessment in writing to that forum or any other international body as far as the Tribunal is aware. Accordingly the Tribunal finds that China has not fulfilled its duties under Article 206 of the [LOS] Convention.[191]

The environmental safeguards policies of MDBs/IFIs (such as the AIIB), typically incorporate *EIA Convention*-type provisions in financing development, which can include those impacting the Third Pole. Whether concerned with hydro-electricity, or other resource or infrastructure development, lenders such as the AIIB will be judged on the effective application of their safeguards policies in this and other Asian subregions. While promoting social advancement and sustainable development,[192] international development places substantial pressure on the

---

[188] Convention on International Trade in Endangered Species of Wild Fauna and Flora (Washington DC, 3 March 1973) 993 UNTS 243, entered into force 1 July 1975 ('CITES').

[189] PCA, above n. 84, para 915, footnote deleted.

[190] Ibid, para 966.

[191] Ibid, para 991, footnote deleted.

[192] See Alex Mourant, Douglas Emeott, Jagabanta Ningthoujam and Jasmin Yu, *Ensuring Sustainability in the Asian Infrastructure Investment Bank and the*

environment; this must be managed very carefully if the Third Pole is to be resilient to them and not be impacted significantly.

Whether funded by the AIIB or other global lenders, development of the China-Pakistan Economic Corridor is a particular concern, in part because of the vast network of road and rail links required to make it a reality. This cross-border route from Kashgar to the port of Gwadar[193] enables China to benefit from the strategic significance of Pakistan to its energy supplies and security, as noted by Niazi at length as follows:

> The port will serve five Chinese ends. First, it will ensure safe shipping for China's energy imports from the Persian Gulf that supplies 60% of its fuel needs. Second, in the event of any hostile action to block its energy supplies through the Persian Gulf, the Gwadar Port will serve as an alternative supply route. Third, it will eventually become a substitute passage for all of Chinese shipments through the Persian Gulf to the Strait of Malacca, where China is totally dependent upon the goodwill of the U.S. and its allies who police it. Fourth, as the Gwadar port sits opposite the Strait of Hurmoz, through which the bulk of the world's energy resources, importantly Japanese fuel imports, are shipped, it will give China a strategic lever to retaliate in case its shipments are obstructed elsewhere. Fifth, above all, the port provides China a strategic foothold in the Arabian Sea and the Indian Ocean. Its presence on the Indian Ocean will further deepen its strategic influence with major South Asian nations, with which its relations are already thriving.[194]

In connection with the AIIB, this was initiated by China in 2014 to provide finance for roads, railways and other infrastructure; the Bank is supported by Asian and non-Asian states and is headquartered in Beijing, emphasizing its importance to China but potentially exposing it if it is not

---

New Development Bank, A SAIS International Development Practicum Report (John Hopkins School of Advanced International Studies, 2015).

[193] For impacts upon the Pakistan town and coastline, see: Jon Boone and Kiyya Baloch, 'A New Shenzhen? Poor Pakistan Fishing Town's Horror at Chinese Plans', *The Guardian*, 4 January 2016. For financial aspects, see: Ali Salman, 'Pakistan-China Economic Corridor: A Cost Benefit Analysis', *The Express Tribune*, 3 May 2015.

[194] Tarique Niazi, 'The Ecology of Strategic Interests: China's Quest for Energy Security from the Indian Ocean and the South China Sea to the Caspian Sea Basin' (2006) 4(4) *China and Eurasian Forum Quarterly* 97, 102. Note the improvement in China's relations with India in recent years as cited elsewhere in this book. Note also that a decade after the publication of Niazi's analysis, China has another potential solution to the access to the Straits of Malacca issue, construction of a canal across Thailand, specifically the less than 30 mile Kra Isthmus. See: Tim Maverick, 'A Thai Canal to be Part of China's Silk Road', *Wall Street Daily*, 27 January 2016.

a success.[195] It is not entirely coincidental that the location of the Asian 'Water Towers' and the origin of the China-Pakistan Economic Corridor are respectively in Tibet and Xinjiang, both the source of major social unrest and a challenge to the territorial integrity of China. Development of these 'Autonomous Provinces' – such as the new rail links to Xinjiang[196] – and the provision of cheap hydro-electric power to Tibet to support new housing and other uses, is part of China's political strategy to integrate these areas into China despite local opposition.[197]

The export of Chinese development to its neighbours is a double-edged sword in managing the escalating social tensions in other countries, in particular between India and Pakistan,[198] and between India and Nepal.[199] In relation to the former, ongoing Chinese support for Pakistan and the routing the Economic Corridor via Kashmir is undoubtedly of concern to India.[200] In relation to the latter, while both India and China have tried to bring Nepal within their sphere of influence following the 2014 Earthquake,[201] Chinese support for Nepal's new Constitution and links with Nepalese Communists (criticized by India) is gaining China the upper hand in its own promotion of Nepalese development. Despite this, there are some indications of China and India acting in concert to

---

195 Richard Javad Heydarian, 'Asian Infrastructure Investment Bank: China's Trojan Horse?', *The Huffington Post*, 21 October 2015; Chi Lo, 'Why AIIB is Important to China's Economic Health', *Hong Kong Economic Journal*, 22 June 2015.

196 See Simon Denyer, 'Bullet Trains Tighten China's Embrace of Restive Xinjiang', *Washington Post*, 10 September 2014; and Agence France-Presse, 'Xinjiang Linked up to High-Speed Rail Network', *South China Morning Post*, 27 December 2014.

197 Emily T Yeh, *Taming Tibet: Landscape Transformation and the Gift of Chinese Development* (Cornell University Press, Ithaca and London, 2013).

198 See Boone and Baloch, above n. 193; and Kiyya Baloch, 'The China-Pakistan Economic Corridor Challenges: One Group is Not Happy about China's Big Investment Plans', *The Diplomat*, 28 September 2015.

199 Sanjoy Majumder, 'Why India is Concerned about Nepal's Constitution', *The Guardian*, 22 September 2015.

200 Conflict in Kashmir is ongoing. See: author unknown, 'Kashmir Clashes over Militant Burhan Wani Leave 30 Dead', *BBC News*, 12 July 2016.

201 Soutik Biswas, 'Nepal Earthquake: How India and China Vie for Influence', *BBC News*, 28 April 2014.

advance economic development and tackle climate change.[202] This provides an important opportunity to improve regulation for environmental protection, which is dependent upon their cooperation.

The relationship between transboundary EIA and the lending policies of MDBs/IFIs has been considered in the literature in the past.[203] More broadly, an example of such an institution involved in the development of international law was the World Bank (WB) in the establishment of the *Indus Waters Treaty*, mentioned above; the WB is a signatory to this agreement and has had a role in the resolution of disputes between India and Pakistan over the allocation of waters.[204] Similarly, the role of the Asian Development Bank (ADB) in the Greater Mekong Subregion (GMS) Programme which led to the *Mekong Agreement* (analysed in Chapter 4), illustrates the broader role of MDBs/IFIs in the development of international law.

As an indication of recent work in this area, which confirms the positive outlook for the *Espoo Convention* in the Asian subregions, the Meeting of Parties (MOP) incorporated a seminar on globalization and the role of MDBs/IFIs as part of the sixth Convention sessions in mid-2014.[205] This brought together government representatives, NGOs, international banks and others 'to provide insights on the practice, needs and expectations of non-ECE countries regarding the implementation of … transboundary environmental impact assessment (EIA) procedures, including in view of their possible future accession to the Convention …'.[206] Focusing on the role of MDBs/IFIs, including on energy-related projects,[207] it engaged

---

202 Debasish Roy Chowdhury, 'Why Modi's India is Warning to China', *South China Morning Post*, 17 September 2015; Reuters, 'China and India Call on Rich Countries to Step up Climate Change Efforts', *The Guardian*, 15 May 2015.

203 Nazari, above n. 148.

204 See *Indus Waters Treaty* 1960, World Bank Responsibilities: <http://web.worldbank.org/WBSITE/EXTERNAL/COUNTRIES/SOUTHASIAEXT/0,,contentMDK:22557065~pagePK:146736~piPK:146830~theSitePK:223547,00.html> accessed 16/11/16.

205 United Nations Economic Commission for Europe, Espoo Convention, Meeting of the Parties Sixth session, Geneva, 2–5 June 2014, Item 6 of the provisional agenda, Seminar on the globalization of the Convention and the Protocol and the role of international financial institutions, Information on the seminar on the globalization of the Convention and the Protocol and the role of international financial institutions, ECE/MP.EIA/2014/INF.8, 28 May.

206 Ibid, para 2.

207 United Nations Economic Commission for Europe, Espoo Convention, Meeting of the Parties, Sixth session, Geneva, 2–5 June 2014, Item 9 of the

with a full and varied range of stakeholders in considering how best to globalize the agreements. It was followed by a draft Declaration proposed by the Bureau[208] on the global application of the Convention and related *SEA Protocol*,[209] and Decision VI/5–II/5 on 'Accession by Member States of the United Nations not members of the United Nations Economic Commission for Europe'.[210]

MDBs/IFIs commonly incorporate environmental safeguards policies as part of their lending practices. The WB comments for example that:

> Environmental Assessment is one of the 10 environmental, social, and legal Safeguard Policies … Environmental Assessment is used … to identify, avoid, and mitigate the potential negative environmental impacts associated with Bank lending operations. In World Bank operations, the purpose of Environmental Assessment is to improve decision making, to ensure that project options under consideration are sound and sustainable, and that potentially affected people have been properly consulted.[211]

An Operational Manual containing the detailed policies was last updated in 2013 and includes transboundary EIA.[212] Other MDBs/IFIs, such as the European Bank for Reconstruction and Development (EBRD) and the

---

provisional agenda, Panel discussion on the application of the Convention and Protocol to energy-related issues, Informal summary of the panel discussion by the moderator.

208 United Nations Economic Commission for Europe, Espoo Convention, Meeting of the Parties, Sixth session, Geneva, 2–5 June 2014, Items 3(d) and 12 of the provisional agenda, Outstanding issues: draft declaration, Adoption of the declaration, Draft declaration proposal by the Bureau, ECE/MP.EIA/2014/L.4–ECE/MP.EIA/SEA/2014/L.4, Part B, 19 March 2014.

209 Protocol on Strategic Environmental Assessment to the Convention on Environmental Impact Assessment in a Transboundary Context (Kiev, 21 May 2003) unreported, in force 11 July 2010.

210 United Nations Economic Commission for Europe, Espoo Convention, Meeting of the Parties, Sixth session, Geneva, 2–5 June 2014, Report of the Meeting of the Parties, addendum, decisions and the declaration adopted jointly by the Meeting of the Parties (Convention and Protocol), ECE/MP.EIA/20/Add.3–ECE/MP.EIA/SEA/4/Add.3, 15 July 2014, 31–32.

211 <http://web.worldbank.org/WBSITE/EXTERNAL/PROJECTS/EXTPOLICIES/EXTSAFEPOL/0,,contentMDK:20543912~menuPK:1286357~pagePK:64168445~piPK:64168309~theSitePK:584435,00.html> accessed 16/11/16.

212 World Bank, *Operational Manual*, OP 4.01 – Environmental Assessment (2003).

ADB, have similar policies;[213] and opportunities for collaboration on strategies, policies, governance, co-financing and other matters have already been canvassed between the EBRD and the AIIB.[214]

Following the establishment of the AIIB in 2014, its own safeguards policies were subsequently drafted.[215] Article 13(4) of its Articles of Association states: 'The Bank shall ensure that each of its operations complies with the Bank's operational and financial policies, including without limitation, policies addressing environmental and social impacts'. In the 'key provisions of the AIIB' on its website it states that these policies will be based on 'international best practices'. China had a key role in the establishment of the Bank,[216] and as it is instrumental in promoting development across Asia and globally, including the Third Pole, the role of the AIIB – which is now open for business – is significant.[217]

Both the AIIB and other MDBs/IFIs, for example the EBRD,[218] clearly suggested that compliance with international law is expected, and their safeguards policies make clear reference to international best practice, including transboundary EIA. Effective implementation is however essential; Chow and Short have considered what the AIIB must do to fulfil its pledge of being 'clean and green'.[219] They cite examples of previous funding by individual Chinese banks, which have neither been transparent nor adhered to expected best practice, as evidence of what to avoid. While they recognize that other MDBs/IFIs have also often not

---

[213] Asian Development Bank, *Environment Safeguards: A Good Practice Sourcebook* (Draft Working Document), December 2012; European Bank for Reconstruction and Development, *Environmental and Social Policy,* 7 May 2014.

[214] Lawrence Chung, 'European Bank to Team Up with AIIB', *Sunday Morning Post*, 28 June 2015.

[215] Asian Infrastructure Investment Bank, *Environmental and Social Framework* (Consultation Draft), 3 August 2015.

[216] Cary Huang and Andrea Chen, 'Deal Seals China's Big Stake in AIIB', *South China Morning Post*, 30 June 2015; Langi Chiang and Man-Ki Kwong, 'Silk Road Deals to Kick off Within Weeks', *South China Morning Post*, 28 February 2015; and Mini Chan, 'China to Create US$40 Billion Silk Road Fund to Upgrade Asia Links' *South China Morning Post*, 9 November 2014.

[217] Glenda Korporaal, 'Asian Infrastructure Bank Open', *The Weekend Australian*, 2–3 January 2016; for detailed discussion of the safeguards, see Marsden, above n. 148.

[218] European Bank for Reconstruction and Development, 2014, 2.

[219] Lowell Chow and Annabel Short, 'Vision in Place', *South China Morning Post*, 27 July 2015.

complied with safeguards policies, they emphasize the significance of the AIIB not failing in this respect. They comment:

> The political context behind the AIIB means scrutiny of its social license to operate will be intense. There is a dire need for improved infrastructure in Asia, but instead of focusing narrowly on the economic benefits the bank can bring to the region, the one biggest factor that will determine the bank's legitimacy and reputation will rest on its social and environmental performance.[220]

This deserves highlighting in connection with the overall focus of this book on the effectiveness of subregional environmental regimes in Asia. This is because the lending policies of MDBs/IFIs are arguably also not only of relevance to these regimes, but in another sense – which is beyond the scope of this book – the operations of the banks themselves may also be regarded as 'regimes' in their own right. As a type of international organization – as seen above in relation to the Mekong and Indus river basins – they sometimes also play significant roles in developing international law as well as expecting compliance from states which benefit from their lending policies. In doing so and in exercising these policies, they also operate across many borders; involve numerous contesting states; impact society, economy and the environment in complex and significant ways; are subject to geopolitical constraints which change frequently as relationships between their member states do; and are framed by laws and institutions of their own, which frequently make reference to international law.

5. CONCLUSIONS, RECOMMENDATIONS AND DIRECTIONS FOR FUTURE RESEARCH

The south and east Asian subregions contain the fewest formal, developed environmental regimes in Asia. With respect to northeast Asia for

---

[220] Ibid. Note also the comments of US President Obama: 'We are all for it … as long as it incorporates strong financial, social and environmental safeguards'. See: Ian Talley, 'Obama: We're All for the Asian Infrastructure Investment Bank', *The Wall Street Journal*, 28 April 2015. Not to be outdone, Russia's President Putin has been a much stronger supporter of integration efforts, particularly in the Russian Far East, although with no mention of the environment. See: Vladimir Putin, 'Family of Nations: The Key to Progress and Development in the Asia-Pacific Region Lies in Ever-Closer Cooperation', *South China Morning Post*, 17 November 2015.

example, Litta comments that 'in contrast to the Southeast Asian experience, regional environmental cooperation in Northeast Asia is much less well developed'.[221] She puts this down to history, different institutional contexts and identities, and how 'The move towards environmental governance in Northeast Asia has been a process of trying to build multilateralism on the back of bilateralism in the absence of a regional institution'.[222] Despite this limitation, the *Tumen Programme* is the only subregional regime across south or east Asia to have made any real progress with transboundary environmental governance. This final section will, in reviewing the findings in relation to south Asia, east Asia and the Third Pole, draw conclusions; it will also make recommendations in relation to the development of a subregional environmental regime for the Third Pole; and will provide suggestions for future research in this area, the impact of which is of significance in many areas of environmental, climate and human rights.

In south Asia, while there are numerous agreements of a bilateral and subregional nature, none are sufficiently well developed to meet the description of a regime laid down in Chapter 1, whether subregional or not. While institutional structures exist to some extent, there is not a close enough link between them and the agreements themselves; environmental objectives and principles are also insufficient, and provisions such as reporting mechanisms and compliance controls are notably absent; finally, practice is non-existent and there are therefore no real opportunities to evaluate effectiveness. As examples, firstly, the *Indus Waters Treaty* does not have an environmental protection focus and involves just India and Pakistan. Secondly, the SAARC agreements, while providing a framework for the breadth of environmental issues, disasters and energy, fail to elaborate detailed obligations which meet the needs of environmental protection and climate resilience. Thirdly, the informal arrangements for air pollution, climate change and marine protection are similarly weak in obligatory nature, failing to proscribe unacceptable behaviour and determine how it is to be dealt with. In south Asia – despite their potential – the subregional institutions are also generally weak and have had a limited role in developing new arrangements or progressing existing ones.

In east Asia as seen in a previous section and noted above, the *Tumen Programme* is the only subregional regime with detailed objectives,

221 Lorraine Elliot, 'East Asia and Sub-Regional Diversity: Initiatives, Institutions and Identity', in L Elliot and S Breslin (eds), *Comparative Environmental Regionalism* (Routledge, Abingdon, UK, 2011) 56, 66.

222 Ibid.

principles and procedures of a transboundary environmental nature, including provisions for transboundary EIA; there is also an institutional structure with regular meetings held, including in the environment and energy areas. However, practice is notably weak in relation to the environment, reflecting the development focus of the arrangements. Evidence of practice is also inhibited by the lack of reporting and compliance provisions in the regime, which perhaps explains the comments of one author noted above.[223] The air pollution arrangements – like the pollutants themselves – are also diffuse; they are also not obligatory, reflective of the difficult geopolitics that positions China alongside Japan and South Korea, with the latter two developed states subject to the pollution from the former undeveloped one. Finally, the regional sea provisions for the East Asian Seas are – even more so than those for the South Asian Seas, which are administered by SACEP – weak and ineffective; this is made worse again by the challenging political situation in this area which is unlikely to improve anytime soon.

In relation to regime effectiveness – a key part of this book which was outlined in detail in Chapter 1 – the conclusions to each of the Chapters 2–4 have for the most part failed to find that the regimes in question have made particularly positive contributions to environmental protection. Part of the problem may lie in the fact that provisions for reporting on environmental performance are either weak or absent; as noted, there is little evidence from south or east Asia to add to this from this chapter. What remains is to address the final two research questions in Chapter 1 that connect the findings of Chapters 2–4 with those in this chapter. The first of these asked whether the institutional arrangements, objectives, principles and provisions were *transferable, or capable of transfer*, here specifically in relation to the Third Pole. The answer to this is generally yes, especially where the other regimes deal with similar matters – such as transboundary watercourses, biodiversity or EIA; however differences of context nonetheless make this a challenging task. The second and final of the research questions asked what China's role was in regard to the subregional regimes – this was answered at the end of each of Chapters 2–4 – and *what can China's role potentially be in the future*, particularly in relation to the Third Pole? This is far more complex given the geopolitics, although some suggestions are made in the recommendations below.

The main recommendation is that a subregional environmental regime should be developed by SAARC and SACEP, building on the existing

[223] Koyano, above n. 101.

agreements in the Asian subregions outlined, and in close cooperation with the international lending banks and China, which should be invited to become a formal member of SAARC and SACEP. The environmental pressures to which the Third Pole is exposed necessitate development of such a regime without delay, because no regulation at all will invariably be worse than regulation which is yet to meet its potential, but can at least guide and provide a framework for sustainable development. Collaboration between China, India, Pakistan and the other states of the subregion, as supported by the lending banks in accordance with environmental and social safeguards policies, is essential to these efforts.

The existing SAARC agreements are foundational to such a regime, potentially shaped by a broader framework. This can potentially be informed by the unratified *Sustainable Development Convention* in central Asia,[224] or the *Conservation Agreement* in southeast Asia,[225] and either may serve as precedents in this respect. However, much more information is needed to account for and understand the advantages and disadvantages that may accrue from experiences elsewhere, as there are clear dangers in failing to recognize the contexts needed for successful legal transplants. The author of this book is particularly minded of the comments made by Alaskan scholar Falk Huettmann, writing in the only scholarly book publication in connection with Third Pole governance to date, who states:

> The last thing we need … are more regulations and laws that do not work and a legal model of 'quick-fixes' that presents bizarre cornucopia and a mosaic of failed policies … Laws without funding and enforcements cannot work; virtually all lawyers know it, yet they do nothing against it, while species increasingly close the distance to their own extinction. It is here where Western society and its culture has failed us so badly again; the United States and its elite law schools are in the lead here. We were fed the idea that more rules, more regulations, and more laws would provide us with a secure and stable fix, making things better (= 'makes sense'). But this concept clearly failed in regard to climate change, human welfare, and in most environmental and wilderness applications.[226]

---

[224] Framework Convention on Environmental Protection for Sustainable Development in Central Asia (Ashgabat, 22 November 2006) TRE-143806, not in force, accessed 20/11/14 at <http://www.ecolex.org/server2.php/libcat/docs/TRE/Full/En/TRE-143806.pdf>.

[225] Agreement on the Conservation of Nature and Natural Resources (Kuala Lumpur, 9 July 1985), not in force.

[226] Falk Huettmann, 'Introduction: Why Three Poles and Why Protect Them', in Falk Huettmann (ed), *Protection of the Three Poles* (Springer, 2012) 3, also quoted in Simon Marsden, 'From the High North to the Roof of the World:

The potential for further treaty building activity, including for biodiversity conservation and EIA specifically, should therefore be considered only after any disadvantages are evaluated alongside any advantages, and with reference to the full range of potential stakeholders – and 'rightsholders'.[227] There is furthermore also major potential to learn from the lessons of relevant intergovernmental forums and organizations elsewhere, for example the Arctic Council, which is itself forging links with other international organizations.[228]

Future research must therefore very carefully analyse possible legal transplants for Third Pole governance. Examples can be found across the MEAs, regional environmental agreements, and subregional environmental agreements in other parts of the world, several, as seen already from examples in this book, with intercontinental (Europe-Asia, Africa-Asia, north America-Europe-north Asia applications). As each of these will have its limitations – in large part because of the context which informs them – elements of good practice must be evaluated and distilled very cautiously in any future regime. In relation to MEAs, the *Biodiversity*[229] and *Climate Change*[230] *Conventions* are of central importance; so too are the other conservation conventions, in particular the *World Heritage*[231] and *Wetlands*[232] *Conventions*, the CMS[233] and CITES.[234] In regard to water governance, which has been seen as central

---

Arctic Precedents for Third Pole Governance' (2016) 8 *The Yearbook of Polar Law* 56–75.

[227] See United Nations Declaration on the Rights of Indigenous Peoples, Resolution adopted by the General Assembly, 61/292, 13 September 2007 (United Nations 2008) ('UNDRIP').

[228] See Marsden, n. 226, and Timo Koivurova, Paula Kankaanpää, and Adam Stępień, 'Innovative Environmental Protection: Lessons from the Arctic' (2015) 27 *Journal of Environmental Law* 285–311.

[229] Above n. 5.

[230] United Nations Framework Convention on Climate Change (Rio de Janiero, 9 May 1992), 1771 UNTS 107, in force 21 March 1994.

[231] Convention Concerning the Protection of the World Cultural and Natural Heritage (Paris, 16 November 1972) 1037 UNTS 151, in force 17 December 1975. Both the World Heritage and Wetlands Conventions enable international legal protection of national sites, which is relevant to the protection of the Third Pole – also comprised of national territory.

[232] Convention on Wetlands of International Importance, especially as Waterfowl Habitat (Ramsar, 2 February 1971) 11 ILM 963, in force 21 December 1975.

[233] Convention on the Conservation of Migratory Species of Wild Animals (Bonn, 23 June 1979) 19 ILM 15, in force 1 November 1983.

[234] Above n. 188.

to many of the agreements across the subregions in this book, both the global UN and regional UNECE *Water Conventions*[235] are of major significance. Key examples are also located in the other polar regions, in particular the *Madrid (Antarctic) Protocol*,[236] the *Svalbard Treaty*[237] and the Arctic Council *EIA Guidelines*.[238] These experiences are also of value in relation to China's role, as it is able to utilize lessons learnt in the Arctic context, for example, for consideration in the Third Pole.[239] The European mountain agreements – the *Alpine*[240] and *Carpathian*[241] *Conventions* – are also of relevance in a sectoral context and should be considered closely, especially for the sustainable development focus of both.

---

235 Above nn. 51 and 53.

236 Protocol on Environmental Protection to the Antarctic Treaty (Madrid, 4 October 1991) 30 ILM 1455 (1991), in force 14 January 1998. Despite what has been said about this being not relevant because of a lack of sustainable development focus, this may change if the Antarctic mining ban ends. For transplant of another kind, see: Simon Marsden, 'Introducing Strategic Environmental Assessment to the Madrid Protocol: Lessons from International Experience' (2011) 1(1) *The Polar Journal* 33–47.

237 Svalbard Treaty (Paris, 9 February 1920) in force 14 August 1925. This is significant because it allows resource exploitation – within environmental protection limits – by states of the international community despite Svalbard being Norwegian territory. Again this has relevance for the development of an international regime in the Third Pole, which is also comprised of separate national territories.

238 Eeva Furman and Mikael Hildén, *Guidelines for Environmental Impact Assessment in the Arctic*, Arctic Environmental Protection Strategy (Finnish Environment Institute, 1997).

239 See Timo Koivurova and Tianbao Qin (eds), *Arctic Law and Governance: The Role of China and Finland* (Hart, Oxford, 2016). Whether based on experiences elsewhere or not, China is not unaware of environmental harm in the Tibetan Plateau; see for example: Edward Wong, 'China Region Bans Glacier Tourism', *International New York Times*, 18 February 2016.

240 Convention on the Protection of the Alps (Salzburg, 7 November 1971) OJ L61, 12/03/1996, p. 32, in force 6 March 1995. See: Volker Mauerhofer, Edwald Galle and Marco Onida, 'The Alps Convention', in CJ Bastmeijer (eds) *Wilderness Protection in Europe: The Role of International, European and National Law* (Cambridge University Press, Cambridge, 2016) 199–221.

241 Framework Convention on the Protection and Sustainable Development of the Carpathians (Kyiv, 22 May 2003) in force January 2006. See: Simon Gaberell, 'Framing Mountains for Regional Environmental Cooperation: The Case of the Carpathians' in Jörg Balsiger and Aysun Uyar (eds), *Comparing Regional Environmental Governance in East Asia and Europe: Proceedings* (Research Institute for Humanity and Nature, Kyoto, 2013) 35–42; and Harald

Aside from these treaty regimes, and as highlighted briefly at the end of the previous section, and in further support of Third Pole regime building, the role of the MDBs/IFIs is worthy of further exploration in a cross-border context. The safeguards policies they apply in their lending practices, and the practical experience of regime building gained in the Mekong and Indus basins, suggest they have much to offer. Beyond this transboundary governance experience, there is also a need for future research to consider carefully the domestic law and policy of the major states in the area, particularly China and India, since transborder regulation should build on national law where possible, as this will already have received government support. While there may not be any clearly identifiable provisions that are relevant to transboundary environmental governance, more in-depth research may indicate a range of potential domestic precedents that can support regime building.

Finally, and in recognition of Huettmann's comments above, consideration also needs to be given to the place of non-traditional legal approaches to assist with Third Pole environmental governance, notably the growing ecological jurisprudence also known as 'Wild Law'.[242] The philosophical approach of this links with the significance of religion in the subregion, in particular Buddhism and Hinduism. Law informed in this way may also assist in regime building, given strong support and practice within society that – where religion is also common across borders – is not constrained by state boundaries. Given the failure of sustainable development to achieve much more than negotiate compromises that have not often put the environment first in decision making, ecological jurisprudence may therefore provide other, additional ways to engage positively with affected people who depend on the environments in question for their livelihoods. Whether concerning the marginalized poor or Indigenous communities – whose traditional relationship with the environment also shares much in common with Wild Law – more effort must hence be put into including, and in the case of Indigenous communities, prioritizing, their involvement in decision making.[243] While

Egerer, Cai Lin Gwee, Eleonora Musco and Mathilde Koeck, 'Wilderness Protection under the Carpathian Convention', in CJ Bastmeijer (eds), *Wilderness Protection in Europe: The Role of International, European and National Law* (Cambridge University Press, Cambridge, 2016) 222–243.

242 Cormac Cullinan, *Wild Law: A Manifesto for Earth Justice* (2nd edition, Green Books, Cambridge, 2011).

243 See UNDRIP above n. 227.

Indigenous issues have not received much attention in this book,[244] there is no question they must be considered in connection with Third Pole governance.[245]

---

[244] For examples see: Janaki Lenin, 'Indigenous Tribe Opposes Hydropower Projects in the Tawang Basin', *The Guardian*, 22 August 2015; and Simon Denyer, 'China Cracks Down on Aggrieved Party Cadres in Xinjiang and Tibet', *Washington Post*, 8 December 2015.

[245] Indigenous issues are a key part of many MEAs, for example the CBD, above n. 5, where Article 8(j) recognizes the role of Indigenous peoples in the conservation and management of biodiversity through the application of Indigenous knowledge. In relation to options for change in the context of the World Heritage Convention (and UNDRIP, above n. 227), see Simon Marsden, 'The World Heritage Convention in the Arctic and Indigenous People: Time to Reform? (2014) 6 *The Yearbook of Polar Law* 226–249; and Simon Marsden, 'The World Heritage Convention: Compliance, Public Participation and the Rights of Indigenous People' (2015) 32(6) *Environmental and Planning Law Journal* 534–547.

Bibliography

Abdulraheem, Mahmoud, Y, 'War-related Damage to the Marine Environment in the ROPME Sea Area', in Jay E Austin and Carl E Bruch (eds), *The Environmental Consequences of War: Legal, Economic, and Scientific Perspectives* (Cambridge University Press, Cambridge, 2007) 338–352

Agence France-Presse, 'South China Sea: US Warns Beijing Against Building "Great Wall of Self-Isolation"', *The Guardian*, 4 June 2016

Agence France-Presse, 'US Military Admits Troops Are Operating Inside Yemen to Combat Al-Qaida', *The Guardian*, 6 May 2016

Agence France-Presse, 'Xinjiang Linked up to High-Speed Rail Network', *South China Morning Post*, 27 December 2014

Agence France-Presse, 'Russia to Build 10 More Nuclear Reactors in India', *South China Morning Post*, 12 December 2014

Agence France-Presse/Jili, 'Air Pollution in Asia Intensifies Cyclones', *The Japan Times*, 1 February 2014

Agyebeng, Kissi, 'Disappearing Acts – Towards a Global Civil Liability Regime for Pollution Damage Resulting from Offshore Oil and Gas Exploration' (2006) *Cornell Law School Graduate Student Papers*, Paper 11

Akiner, Shirin, 'Ten Years On: Achievements, Prospects, New Concerns', in Shirin Akiner (ed), *The Caspian: Politics, Energy and Security* (Routledge, Abingdon, UK and New York, USA, 2004) 337

Al-Abdulrazzak, Dalal, Dirk Zeller, Dyhia Belhabib, Dawit Tesfamichael, Daniel Pauly and Al-Awadi, 'Total Marine Fisheries Catches in the Persian/Arabian Gulf from 1950 to 2010' (2015) 2 *Regional Studies in Marine Science* 28–34

Al-Awadi, Abdulrahman, *The Environmental Catastrophe in Kuwait* (ROPME, undated)

Al-Hussan, Jassim Mohammad, *The Iraqi Invasion of Kuwait, an Environmental Catastrophe* (Jassim Mohammad Al-Hussan, 1992)

Amineh, M Parvizi (ed), *The Greater Middle East in Global Politics: Social Science Perspectives on the Changing Geography of the World Politics* (Brill, Leiden, 2007)

Arsel, Murat and Max Spoor, *Water, Environmental Security and Sustainable Development: Conflict and Cooperation in Central Eurasia* (London, Routledge, 2010)

Ascher, William and Natalia Mirovitskaya, *The Caspian Sea: A Quest for Environmental Security* (Kluwer, Dortrecht, 2000)

Asian Development Bank, Environment Division, *Central Asia Environments in Transition* (Manila, 1998)

Asian Development Bank, *Emerging Asian Regionalism: A Partnership for Shared Prosperity* (Manila, 2008)

Asian Development Bank, *Environment Safeguards: A Good Practice Sourcebook* (Draft Working Document), December 2012

Asian Infrastructure Investment Bank, *Environmental and Social Framework* (Consultation Draft), 3 August 2015

Attard, David, Malgosia Fitzmaurice, Norman Martinez and Riyaz Hamza (eds), *The IMLI Manual on International Maritime Law. Vol. III, Marine Environmental Law and International Maritime Security Law* (Oxford, Oxford University Press, 2016)

Australian Associated Press, 'Mount Everest Melting, Chinese Study Says', *The Mercury*, 8 December 2015

Author unknown, 'Kashmir Clashes over Militant Burhan Wani Leave 30 Dead', *BBC News*, 12 July 2016

Author unknown, 'Saudi Arabia and Egypt Announce Red Sea Bridge', *BBC News*, 8 April 2016

Author unknown, 'US$10 Bln in Deals Signed as Modi Visits Beijing', *Hong Kong Economic Journal*, 15 May 2015

Bach, Hanne, Jeremy Bird, Torkil Jønch Clausen, Kurt Mørck Jensen, Rane Baadsgaard Lange, Robin Taylor, Vitoon Viriyasakultorn and Aaron Wolf, *Transboundary River Basin Management: Addressing Water, Energy and Food Security* (Mekong River Commission, Vientiane, 2012)

Bach, Hanne, Paul Glennie, Robin Taylor, Torkil Clausen, Fritz Holzwarth, Kurt Mørck Jensen, Abel Meija and Susanne Schmeier, *Cooperation for Water, Energy, and Food Security in Transboundary Basins under Changing Climate* (Mekong River Commission, Vientiane, 2014)

Badenoch, Nathan, *Transboundary Environmental Governance: Principles and Practice in Mainland Southeast Asia* (World Resources Institute, 2002)

Ballaman, Richard, 'Comparing Regional Environmental Governance in East Asia and Europe – Air Pollution', in Jörg Balsiger and Aysun Uyar (eds), *Comparing Regional Environmental Governance in East Asia and Europe: Proceedings* (Research Institute for Humanity and Nature, Kyoto, 2013) 71–74

Baloch, Kiyya, 'The China-Pakistan Economic Corridor Challenges: One Group is Not Happy about China's Big Investment Plans', *The Diplomat*, 28 September 2015

Balsiger, Jörg and Aysun Uyar, 'Introduction: Comparing Regional Environmental Governance in East Asia and Europe', in Jörg Balsiger and Aysun Uyar (eds), *Comparing Regional Environmental Governance in East Asia and Europe: Proceedings* (Research Institute for Humanity and Nature, Kyoto, 2013) 1–6

Balsiger, Jörg and Aysun Uyar (eds), *Comparing Regional Environmental Governance in East Asia and Europe: Proceedings* (Research Institute for Humanity and Nature, Kyoto, 2013)

Bao, Cun-kuan, Lu Yong-sen and Shang Jin-cheng, 'Framework and Operational Procedure for Implementing Strategic Environmental Assessment in China' (2004) 24 *Environmental Impact Assessment Review* 27–46

Bart, Jason, 'Weaponizing Water: Water and Energy as Sources of Conflict among the Central Asian Soviet Successor States' (2013) 22(1) *Michigan State International Law Review* 409

Bastmeijer, Kees and Timo Koivurova (eds), *Theory and Practice of Transboundary EIA* (Martinus Nijhoff, 2008)

Beeson, Mark, 'The Coming of Environmental Authoritarianism' (2010) 19(2) *Environmental Politics* 276–294

Best, Edward, 'The Assessment of Regional Governance: Principles, Indicators and Potential Pitfalls', UNU-CRIS Working Papers, W-2008/10 (United Nations University, 2008) 1–27

Best, Edward, 'Regional Integration and (Good) Regional Governance: Are Common Standards and Indicators Possible?', UNU-CRIS Working Papers, O-2004/6 (United Nations University, 2004) 1–45

Bichsel, Christine, 'Liquid Challenges: Contested Water in Central Asia' (2011) 12(1) *Sustainable Development Law and Policy* 24.

Bina, Olivia, 'Context and Systems: Thinking More Broadly about Effectiveness in Strategic Environmental Assessment in China' (2008) 42 *Environmental Management* 717–733

Biswas, Soutik, 'Nepal Earthquake: How India and China Vie for Influence', *BBC News*, 28 April 2014

Black, A, H Esmaeili and N Hosen, *Modern Perspectives on Islamic Law* (Edward Elgar, Cheltenham, 2013)

Blanchard, Jean-Marc F, 'Giving the Unrecognized their Due: Regional Actors, International Institutions and Multilateral Economic Cooperation in Northeast Asia', in DW Drezner (ed), *Locating the Proper Authorities: The Interaction of Domestic and International Institutions* (Ann Arbor, University of Michigan Press, 2003) 49–76

Blanchard, Jean-Marc F, 'The Heyday of Beijing's Participation in the Tumen River Area Development Programme, 1990–95: A Political Explanation' (2000) 9 *Journal of Contemporary China* 271–290

Bliss, Andrew, Regine Hock and Valentina Radić, 'Global Response of Glacier Runoff to Twenty-First Century Climate Change' (2014) 119 *Journal of Geophysical Research: Earth Surface* 717–730

Boer, B, P Hirsch, S Johns, B Saul and N Scurrah, *The Mekong: A Socio-legal Approach to River Basin Development* (Routledge, Abingdon, UK, 2015)

Boer, B, R Ramsar and DR Rothwell, *Environmental Law in the Asia Pacific* (Kluwer, London, 1998)

Boilard, Marie Christine, 'Improving Policy Responses to Piracy in the Red Sea and Gulf of Aden: What Role for India?' (2010) 34(4) *Strategic Analysis* 625–638

Boone, Jon, 'Afghanistan and Pakistan Earthquake Death Toll Will Rise, Say Officials', *The Guardian*, 27 October 2015

Boone, Jon and Kiyya Baloch, 'A New Shenzhen? Poor Pakistan Fishing Town's Horror at Chinese Plans', *The Guardian*, 4 January 2016

Borger, Julian, 'South China Sea Fears Grow Before Tribunal Rules on Disputed Islands', *The Guardian*, 29 May 2016

Bosworth, William, Philippe Huchon and Ken McClay, 'The Red Sea and Gulf of Aden Basins' (2005) 43 *Journal of African Earth Sciences* 334–378

Boyle, Alan, 'Developments in the International Law of Environmental Impact Assessments and their Relation to the Espoo Convention' (2011) 20(3) *Review of European Community and International Environmental Law* 227–231

Breau, Susan, 'Protection of the Environment during Armed Conflict', in S Alam, MJH Bhuiyan, T Chowdhury and E Techera (eds), *Routledge Handbook of International Environmental Law* (Routledge, London, 2013) 617–632

Breitmeir, H, OR Young and M Zurn, *Analysing International Environmental Regimes: From Case Study to Database* (MIT Press, Cambridge, 2006)

Breu, Thomas, Hans Hurni, Andrea Wurth Stucki, *The Tajik Pamirs: Challenges of Sustainable Development in an Isolated Mountain Region* (Centre for Development and Environment, University of Berne, Berne, 2003)

Brooks, Irene B and Liu Hongxia, 'River Basin Commissions as a Mechanism for Mitigating and Resolving Conflicts' (2006) *China Environment Series* 202–208

Bruch, Carl and Elizabeth Mrema, *Manual on Compliance with and Enforcement of Multilateral Environmental Agreements* (UNEP, Nairobi, 2006)

Bruch, Carl, Mikiyasu Nakayama, Jessica Troell, Lisa Goldman and Elizabeth Maruma Mremaet, 'Assessing the Assessments: Improving

Methodologies for Impact Assessment in Transboundary Watercourses' (2008) 26 *Impact Assessment and Project Appraisal* 241–242
Burke, J, 'India-China Border Standoff Highlights Tensions Before Xi Visit', *The Guardian*, 16 September 2014
Buxton, Charles, 'NGO Networks in Central Asia and Global Civil Society: Potentials and Limitations' (2009) 28(1) *Central Asian Survey* 43–58
Byers, Michael, *International Law and the Arctic* (Cambridge University Press, Cambridge, 2015)
Byrne, Adam, 'The 1979 Convention on Long-Range Transboundary Air Pollution: Assessing its Effectiveness as a Multilateral Environmental Regime after 35 Years' (2015) 4(1) *Transnational Environmental Law* 37–67
Cabestan, Jean-Pierre, 'Energy Cooperation between China and Central Asia' (European Council on Foreign Relations/Asia Centre, 2011) 6–8
Callus, Andrew and Stephen Jewkes, 'Crippled Kashagan Oil Project a Bureaucratic "Nightmare"', *Reuters*, 2 January 2014
Campins Eritja, Mar, Jaume Saura Estapá and Xavier Pons Ráfols, 'Towards Improved Regional Co-operation over Water Uses in Central Asia: The Case of Hydroelectric Energy and Inland Fisheries' (2016) 6(1) *Asian Journal of International Law* 119–158
Carrell, Severin and Libby Brooks, 'Nicola Sturgeon: Second Scottish Independence Poll Highly Likely', *The Guardian*, 25 June 2016
Carrera, Alexandra, 'Kashagan: A Look at Caspian Sea Oil and Natural Gas Production', 23 September 2013
Carty, Anthony and Fozia Nazir Lone, 'Some New Haven International Law Reflections on China, India and their Various Territorial Disputes' (2011) 19(1) *Asia Pacific Law Rev*iew 93–112
Cassar Angela Z and Carl E Bruch, 'Transboundary Environmental Impact Assessment in International Watercourse Management' (2003) 12 *New York University Environmental Law Journal* 194–195
Cavaleri, Emanuela Maria, 'The Post-2015 UN Development Framework: Perspectives for Regional Involvement', UNU-CRIS Working Papers, W-2014/12 (United Nations University, 2012) 1–28
Chan, Gerald, 'China's Compliance in Global Environmental Affairs' (2004) 45(1) *Asia Pacific Viewpoint* 69–86
Chan, Mini, 'China to Create US$40 Billion Silk Road Fund to Upgrade Asia Links', *South China Morning Post*, 9 November 2014
Chan, Mini, 'Agreements Reached on a Number of Issues, But No Progress on Long-Running Border Row', *South China Morning Post*, 21 September 2014
Chen, Stephen, 'Skin-Deep Explanation for Tibet's Bulging Plateau', *South China Morning Post*, 24 January 2015

Chesterman, Simon, 'The International Court of Justice in Asia: Interpreting the *Temple of Preah Vihear* Case' (2015) 5(1) *Asian Journal of International Law* 1–6

Chettri, N, AB Shrestha, Y Zhaoli, B Bajracharya, E Sharma and H Ouyang, '"Real World" Protection for the Third Pole and its People', in Falk Huettmann (ed), *Protection of the Three Poles* (Springer, Japan, 2012) 113–132

Chiang, Langi and Kwong, Man-Ki, 'Silk Road Deals to Kick off Within Weeks', *South China Morning Post*, 28 February 2015

Chow, Lowell and Annabel Short, 'Vision in Place', *South China Morning Post*, 27 July 2015

Chowdhury, Debasish Roy, 'Why Modi's India is Warming to China', *South China Morning Post*, 17 September 2014

Chul, Ryu Ji, *GTI Energy Cooperation Programme in Northeast Asia* (Greater Tumen Initiative, February 2016)

Chung, Lawrence, 'European Bank to Team up with AIIB', *Sunday Morning Post*, 28 June 2015

Chung, Lawrence, 'Chinese President Xi Jinping Visits Kazakhstan, Russia and Belarus to Discuss Silk Road Belt', *Sunday Morning Post*, 8 May 2015

Clarke, Michael, 'China and the Shanghai Cooperation Organization: The Dynamics of "New Regionalism", "Vassalization", and Geopolitics in Central Asia', in Emilian Kavalski (ed), *The New Central Asia: The Regional Impact of International Actors* (World Scientific, Singapore, 2010) 117–148

Clough, Ralph N, 'The Status of Taiwan in the New International Legal Order in the Western Pacific' (1993) 87 *Proceedings of the Annual Meeting (American Society of International Law)*, Challenges to International Governance, 73–77

Contreras, AP, 'Transboundary Environmental Governance in Southeast Asia', in A Pandya and E Laipson (eds), *Transnational Trends: Middle Eastern and Asian Views* (The Henry L Stimson Center, July 2008) 155–170

Cooke, Kieran, 'The Dams of India: Boon or Bane?', *The Guardian*, 17 March 2014

Coskun, Bezen Balamir, 'Region and Region Building in the Middle East: Problems and Prospects', UNU-CRIS Working Papers, O-2006/1 (United Nations University, 2006) 1–17

Craik, Neil, *The International Law of Environmental Impact Assessment: Process, Substance and Integration* (Cambridge University Press, Cambridge, 2008)

Cullet, Philippe and Joyeeta Gupta, 'India: Evolution of Water Law and Policy', in Joseph W Dellapenna and Joyeeta Gupta (eds), *The Evolution of the Law and Politics of Water* (Springer, 2009) 157–173

Cullinan, Cormac, *Wild Law: A Manifesto for Earth Justice* (2nd edition, Green Books, Cambridge, 2011)

Daming HE, Zhao Wenjuan and Feng Yan, 'Research Progress of International Rivers in China' (2004) 14 *Journal of Geographical Sciences* 22

Daniel, JR, SL Pinel and J Brooks, 'Overcoming Barriers to Transboundary Water Governance' (2013) 33(3) *Mountain Research and Development* 215–224

Davies, Ian and Valery Neronov, 'Field Diagnostic and Planning Mission to the Lower Tumen River Area of China and the Russian Federation', 26 May–7 June 2003

Denyer, Simon, 'China Cracks Down on Aggrieved Party Cadres in Xinjiang and Tibet', *Washington Post*, 8 December 2015

Denyer, Simon, 'Bullet Trains Tighten China's Embrace of Restive Xinjiang', *Washington Post*, 10 September 2014

De Mora, Stephen J and Tim Turner, 'The Caspian Sea: A Microcosm for Environmental Science and International Cooperation' (2004) 48 *Marine Pollution Bulletin* 26–29

Desrina, R, Chairil Anwar and Tri Muji Susantoro, 'Environmental Impacts of the Oil and Gas Platform Decommissioning' (2013) 36 *Scientific Contributions Oil and Gas* 98

Dinar, Ariel, S Dinar, S McCaffrey and D Mckinney, *Bridges over Waters: Understanding Transboundary Water Conflict, Negotiation and Cooperation* (World Scientific, Singapore, 2013)

Dore, John, 'The Governance of Increasing Mekong Regionalism', paper presented to Regional Centre for Social and Sustainable Development conference, Chiang Mai University, 11–14 July 2003

Dunlap, Ben N, 'Divide and Conquer? The Russian Plan for Ownership of the Caspian Sea' (2004) 27(1) *Boston College International and Comparative Law Review* 115–130

Duran, Hasan and Kadir Kürşat Yilmaz, 'Energy', in EAH Gencer and C Gerni (eds), *Central Asia Economies in Transition* (Cambridge Scholars, Newcastle, 2012)

Duyck, Sébastien, *Legal Issues Related to Options for a Regional-Seas-Type Arrangement for the Arctic Ocean*, Northern Institute for Environmental and Minority Law

Earle, A, A Jagerskog and J Öjendal (eds), *Transboundary Water Management: Principles and Practice* (Earthscan, London, 2010)

Ebbesson, Jonas, 'Principle 10: Public Participation', in Jorge E Viñuales (ed), *The Rio Declaration on Environment and Development: A Commentary* (Oxford University Press, Oxford, 2015)

Ebbesson, Jonas, 'Planetary Boundaries and the Matching of International Treaty Regimes' (2014) 59 *Scandinavian Studies in Law* 260–284

Egerer, Harald, Cai Lin Gwee, Eleonora Musco and Mathilde Koeck, 'Wilderness Protection under the Carpathian Convention', in CJ Bastmeijer (ed), *Wilderness Protection in Europe: The Role of International, European and National Law* (Cambridge University Press, Cambridge, 2016) 222–243

Elliott, Lorraine, 'ASEAN and Environmental Governance: Rethinking Networked Regionalism in Southeast Asia' (2011) 14 *Procedia – Social and Behavioural Sciences* 61–64.

Elliott, Lorraine, 'East Asia and Sub-Regional Diversity: Initiatives, Institutions and Identity', in L Elliott and S Breslin (eds), *Comparative Environmental Regionalism* (Routledge, Abingdon, UK, 2011) 56–75

Erlewein, Alexander, 'Disappearing Rivers – The Limits of Environmental Assessment for Hydropower in India' (2013) 43 *Environmental Impact Assessment Review* 135–143

Esmaeili, Hossein, *The Legal Regime of Offshore Oil Rigs in International Law* (Ashgate Dartmouth, 2001)

Esmaeili, Hossein and Brendan Grigg, 'Pollution from Dumping', in David Attard, Malgosia Fitzmaurice, Norman Martinez and Riyaz Hamza (eds), *The IMLI Manual on International Maritime Law. Vol. III, Marine Environmental Law and International Maritime Security Law* (Oxford University Press, Oxford, 2016)

Esty, DC, 'Breaking the Environmental Law Logjam: The International Dimension' (2008) 17(1) *N.Y.U. Environmental Law Journal* 836–853European Bank for Reconstruction and Development, *Environmental and Social Policy*, 7 May 2014

Eyler, Brian, 'China Needs to Change its Energy Strategy in the Mekong Region', in China Dialogue, *Forging a New Course for the Mekong* 14–16

Faure, Michael and Song Ying (eds), *China and International Environmental Liability: Legal Remedies for Transboundary Pollution* (Edward Elgar, Cheltenham, 2008)

Fawcett, Louise, 'The History and Concept of Regionalism', UNU-CRIS Working Papers W-2013/5 (United Nations University, 2013) 1–19

Féaux de la Croix, Jeanne, 'Moving Metaphors We Live By: Water and Flow in the Social Sciences and Around Hydroelectric Dams in Kyrgyzstan' (2011) 30(3–4) *Central Asian Survey* 487–502

Fioramonti, Lorenzo, 'The Evolution of Supranational Regionalism: From Top-down Regulatory Governance to Sustainability Regions?', UNU-CRIS Working Papers W-2014/2 (United Nations University, Brugge, 2014) 1–26

Fischer, Thomas (ed), *Environmental Assessment: Critical Concepts in Built Environment. Vol. 4, Environmental Assessment Effectiveness* (Routledge, Abingdon, 2015)

Ford, John, '*Philippines v. China* Won't End South China Sea Disputes', *The Diplomat*, 17 March 2016

Framework Convention for the Protection of the Marine Environment of the Caspian Sea, Fifth Meeting of the Conference of the Parties, *Final Act*, 30 May 2014, Ashgabat

Framework Convention for the Protection of the Marine Environment of the Caspian Sea, Conference of the Parties Fifth Meeting [29–30 May 2014, Ashgabat] Turkmenistan, Item 4 of the Provisional Agenda, *Implementation of the Framework Convention for the Protection of the Marine Environment of the Caspian Sea* (Tehran Convention), Report by the Interim Secretariat TC/COP5/2 ed. 15 February 2014

Framework Convention for the Protection of the Marine Environment of the Caspian Sea, Conference of the Parties Fifth Meeting [29–30 May 2014, Ashgabat] Turkmenistan, Items 5 and 12 of the Provisional Agenda, *Recommendations and Analysis Based on the Comments by the Aarhus Convention Secretariat on Public Participation under the Tehran Convention*, Note by the Interim Secretariat, TC/COP5/INF.6

Framework Convention for the Protection of the Marine Environment of the Caspian Sea, Conference of the Parties Fifth Meeting [29–30 May 2014, Ashgabat] Turkmenistan, Items 5 and 12 of the Provisional Agenda, *Framework for the Implementation of the Caspian Environmental Monitoring Programme (EMP)*, Note by the Interim Secretariat, TC/COP5/5, 6 March 2014

Fumagalli, Matteo, 'The United States and Central Asia', in E Kavalski, *The New Central Asia: The Regional Impact of International Actors* (World Scientific, Singapore, 2010) 177–190

Furman, Eeva and Mikael Hildén, *Guidelines for Environmental Impact Assessment in the Arctic*, Arctic Environmental Protection Strategy (Finnish Environment Institute, 1997)

Gaberell, Simon, 'Framing Mountains for Regional Environmental Cooperation: The Case of the Carpathians', in Jörg Balsiger and Aysun Uyar (eds), *Comparing Regional Environmental Governance in East Asia and Europe: Proceedings* (Research Institute for Humanity and Nature, Kyoto, 2013) 35–42

Gan, Nectar and Zhen Liu, 'Xi Jinping Holds Out His Hand to Washington as China-US Talks Begin', *South China Morning Post*, 6 June 2016

Garschagen, Matthias, Javier Revilla Diez, Dang Kieu Nhan and Frauke Kraas, 'Socio-Economic Development in the Mekong Delta: Between the Prospects for Progress and the Realms of Reality', in FG Renaud and C Kuenzer, *The Mekong Delta System: Interdisciplinary Analyses of a River Delta* (Springer, 2012) 83–132

Gault, Simon, Steven Hazelwood, Andrew Tettenborn, Stephen Girvin, Stewart Buckingham, Thomas Macey-Dare and Edward Cole (eds), *Marsden and Gault on Collisions at Sea* (14th edition, Sweet and Maxwell, London, 2016)

Gerges, Makram A, 'The Red Sea and Gulf of Aden Action Plan – Facing the Challenges of an Ocean Gateway' (2002) 45 *Ocean and Coastal Management* 885–903

Gerlak, Andrea K, 'One Basin at a Time: The Global Environment Facility and Governance of Transboundary Waters' (2004) 4(4) *Global Environmental Politics* 108–141

Ghabra, Shafeeq, 'Between Iran and the Gulf: An Environment in Transition', *The Washington Institute for Near East Policy*, 16 May 2016

Gladstone, William, 'Editorial: Towards Conservation of a Globally Significant Ecosystem: The Red Sea and Gulf of Aden' (2008) 18 *Aquatic Conservation: Marine and Freshwater Ecosystems* 1–5

Gladstone, William, Fareed Kruup and Mohammed Younis, 'Development and Management of a Network of Marine Protected Areas in the Red Sea and Gulf of Aden Region' (2003) 46 *Ocean and Coastal Management* 741–761

Gladstone, William, N Tawfiq, D Nasr, I Andersen, C Cheung, H Drammeh, F Krupp and S Lintner, 'Sustainable Use of Renewable Resources and Conservation in the Red Sea and Gulf of Aden: Issues, Needs and Strategic Actions' (1999) 42 *Ocean and Coastal Management* 671–697

Glantz, M, *Creeping Environmental Problems and Sustainable Development in the Aral Sea Basin* (Cambridge University Press, Cambridge, 1999)

Glenn, H Patrick, *Legal Traditions of the World: Sustainable Diversity in Law* (5th edition, Oxford University Press, Oxford, 2014)

Gochhayat, A, 'Regionalism and Sub-Regionalism: A Theoretical Framework with Special Reference to India' (2014) 8(1) *African Journal of Political Science and International Relations* 10–26

Godement, François, 'Introduction', in European Council on Foreign Relations, *The New Great Game in Central Asia*, China Analysis/Asia Centre, September 2011

Goldenberg, Suzanne, 'Plans for Coal-Fired Power in Asia are "Disaster for Planet" Warns World Bank', *The Guardian*, 5 May 2016

Goodsite, Michael, Catherine Wong, Ole Hertel and Lars Moseholm, 'Responses to Air Pollution Based on Historical and Current Policies in the EU and ASEAN' (2011) 6 *Global Environment* 150–182

Greater Tumen Initiative, 13th GTI Consultative Commission Meeting, *Greater Tumen Initiative Strategic Action Plan for the Years 2012–2015*, 10 October 2012

Green, James A, *The Persistent Objector Rule in International Law* (Oxford University Press, Oxford, 2016)

Haidah, Suhasini and Damakant Jayshi, 'Modi Talks About "Closer South Asian Integration" at SAARC Summit', *The Hindi*, 26 November 2014

Haidah, Suhasini and Damakant Jayshi, 'SAARC Leaders Reach Last Minute Energy Deal', *The Hindi*, 27 November 2014

Harvey, Fiona, 'Brexit Would Damage UK Environment, Say Experts', *The Guardian*, 27 January 2016

Hensengerth, Oliver, 'Where Is the Power? Transnational Networks, Authority and the Dispute over the Xayaburi Dam on the Lower Mekong Mainstream' (2015) 40(5–6) *Water International* 911–928

Hensengerth, Oliver, 'Global Norms in Domestic Politics: Environmental Norm Contestation in Cambodia's Hydropower Sector' (2015) 28(4) *The Pacific Review* 505–528

Hensengerth, Oliver, 'Water Governance in the Mekong Basin: Scalar Tradeoffs, Transnational Norms and Chinese Hydropower Investment', in P Nyíri and D Tan (eds), *How Chinese Engagements are Changing Southeast Asia: People, Money, Ideas and their Effects* (University of Washington Press, Seattle, 2015)

Hewitt, K, *Glaciers of the Karakoram Himalaya: Glacial Environments, Processes, Hazards and Resources* (Springer, 2014)

Heydarian, Richard, 'Asian Infrastructure Investment Bank: China's Trojan Horse?', *The Huffington Post*, 21 October 2015

Hijioka, Y, Lin, E, Pereira, JJ, Corlett, RT, Cui, X, Insarov, GE, Lasco, RD, Lindgren, E and Surjan, A, 'Asia', in VR Barros et al., *Climate Change 2014: Impacts, Adaptation and Vulnerability. Part B: Regional Aspects. Contribution of Working Group II to the Fifth Assessment Report of the Intergovernmental Panel on Climate Change* (Cambridge University Press, Cambridge, 2014) 1327–1370

Hill, Joe, 'The Role of Authority in the Collective Management of Hill Irrigation Systems in the Alai (Kyrgyzstan) and Pamir (Tajikistan)' (2013) 33(3) *Mountain Research and Development* 294–304

Hilton, Isabel, 'Introduction', in *Brahmaputra: Towards Unity* (thethirdpole.net, undated)

Hinrichsen, Don, *Coastal Waters of the World: Trends, Threats, and Strategies* (Island Press, Washington DC, 1998)

Hirsch, Philip, 'Laos Mutes Opposition to Controversial Mekong Dam', in China Dialogue, *Forging a New Course for the Mekong* 36–38

Hirsch, Philip and Kurt Mørck Jensen with Ben Boer, Naomi Carrard, Stephen FitzGerald and Rosemary Lyster, *National Interests and Transboundary Water Governance in the Mekong* (Australian Mekong Resource Centre, University of Sydney, Danish International Development Assistance, 2006)

Horberry, J, 'Transboundary Environmental Assessment: The Experience of the Mekong River Commission', presentation to the third MOP to the Espoo Convention, 2004

Horton, Chris, 'Treading Warily on China', *International New York Times*, 26 May 2016

Huang, Cary and Andrea Chen, 'Deal Seals China's Big Stake in AIIB', *South China Morning Post*, 30 June 2015

Huettmann, Falk, 'Introduction: Why Three Poles and Why Protect Them', in Falk Huettmann (ed), *Protection of the Three Poles* (Springer, Japan, 2012) 3–32

Hunter, David B, *International Financial Institutions and International Law* (Kluwer Law International, the Netherlands, 2010)

International Center for Environmental Management, *MRC Strategic Environmental Assessment (SEA) of Hydropower on the Mekong Mainstream: Summary of the Final Report* (Mekong River Commission, Hanoi, 2010)

International Center for Environmental Management, *MRC Strategic Environmental Assessment of Hydropower on the Mekong Mainstream: Final Report* (Mekong River Commission, Hanoi, 2010)

Interstate Sustainable Development Commission/Regional Mountain Centre for Central Asia, *Strategy of the Regional Mountain Centre for Central Asia*, Bishkek, 2009

Ip, Eric C, 'The Power of International Legal Personality in Regional Integration', UNU-CRIS Working Papers W-2010/4 (United Nations University, Brugge, 2010) 1–22

Ives, Mike, 'Climate Change Could Worsen Mekong Delta Woes', in China Dialogue, *Forging a New Course for the Mekong* 41–43

Jacobsson, Mans, 'Liability and Compensation for Ship Source Pollution', in David Attard, Malgosia Fitzmaurice, Norman Martinez and Riyaz Hamza (eds), *The IMLI Manual on International Maritime Law. Vol. III, Marine Environmental Law and International Maritime Security Law* (Oxford University Press, Oxford, 2016)

Janusz, Barbara, 'The Framework Convention for the Protection of the Marine Environment of the Caspian Sea' (2005) 4(1) *Chinese Journal of International Law* 257–270

Jayakumar, S, Tommy Koh, Robert Beckman and Hao Duy Phan (eds), *Transboundary Pollution: Evolving Issues of International Law and Policy* (Edward Elgar, Cheltenham, 2015)

Jerger, David B, 'Indonesia's Role in Realizing the Goals of ASEAN's Agreement on Transboundary Haze Pollution' (2014) 14(1) *Sustainable Development Law and Policy* 35–74

Johns, Fleurs, Ben Saul, Philip Hirsch, Tim Stephens and Ben Boer, 'Law and the Mekong River Basin: A Socio-Legal Research Agenda on the Role of Hard and Soft Law in Regulating Transboundary Water Resources' (2010) 11(1) *Melbourne Journal of International Law*, 154–174

Johnson, David, Angela Benn and Maria Adelaide Ferreira, *Measuring Success: Indicators for the Regional Seas Conventions and Action Plans* (United Nations Environment Programme, 2014)

Johnson, Rob, *Oil, Islam and Conflict: Central Asia since 1945* (Reaktion Books, London, 2007)

Jones, Gavin W, *The Population of Southeast Asia* (Asia Research Institute, Working Paper Series 196, National University of Singapore, 2013)

Kalra, Aditya and Asad Hashim, 'India-Pakistan Peace Talks Collapse, Deadlock Sours Relations', *Reuters*, 23 August 2015

Kandiyoti, Rafael, 'What Price Access to the Open Seas? The Geopolitics of Oil and Gas Transmission from the Trans-Caspian Republics' (2008) 27(1) *Central Asian Survey* 75–93

Kanie, N, S Andresen and PM Haas (eds), *Improving Global Environmental Governance: Best Practices for Architecture and Agency* (Routledge, Abingdon, 2014)

Karrar, Hasan H, *The New Silk Road Diplomacy: China's Central Asian Foreign Policy since the Cold War* (UBC Press, Vancouver, 2010)

Kassam, Ashifa and Tom Phillips, 'Chinese Minister Vents Anger When Canadian Reporter Asks About Human Rights', *The Guardian*, 2 June 2016

Kato, K and W Takahashi, *Regional/Subregional Environmental Cooperation in Asia* (IGES, 2001)

Kavalski, Emilian, 'Uncovering the "New" Central Asia: The Dynamics of External Agency in a Turbulent Region', in Emilian Kavalski (ed), *The New Central Asia: The Regional Impact of International Actors* (World Scientific, Singapore, 2010) 1–25

Kazantsev, Andrei, 'Russian Policy in Central Asia and Caspian Sea Region' (2008) 20(6) *Europe-Asia Studies* 1073–1088

Kazu Kato, 'An Analytical Framework for a Comparative Study of Subregional Environmental Programmes in Asia', in K Kato and

W Takahashi, *Regional/Subregional Environmental Cooperation in Asia* (IGES, 2001) 5

Kelly, Katherine M, 'Declaring War on the Environment: The Failure of International Environmental Treaties During the Persian Gulf War' (1992) 7(4) *American University International Law Review* 921–950

Kepbanov, Yolbars A, 'The New Legal Status of the Caspian Sea is the Basis of Regional Cooperation and Stability' (1997–1998) II(4) *Perceptions – Journal of International Affairs* 5pp

Khan, Nuzrat Yar and Dhari Al-Ajmi, 'Post War Imperatives for the Sustainable Management of the Gulf Eco-System' (1998) 24 (1–2) *Environment International* 239–248

Kim, Inkyoung, 'Environmental Cooperation of Northeast Asia: Transboundary Air Pollution', (2007) 7 *International Relations of the Asia-Pacific* 439–462

Kim, Inkyoung, 'Messages from a Middle Power: Participation by the Republic of Korea in Regional Environmental Cooperation on Transboundary Air Pollution Issues' (2014) 14 *International Environmental Agreements* 147–162

Kim, Juli S, 'Transboundary Air Pollution – Will China Choke on its Success?', *A China Environmental Health Project Fact Sheet*

Kim, Myungjin, 'Environmental Cooperation in Northeast Asia' (2004) 22 *Impact Assessment and Project Appraisal* 191–203

Kimberley, Neal, 'Lower Mekong's Water Woes Mainly Due to El Nino, But China still Damned for Dams', *South China Morning Post*, 3 May 2016

King, Hugh and Laurence Smith, 'Many Rivers to Cross: Evaluating the Benefits and Limitations of Strategic Environmental Assessment for the Koshi River Basin' (2016) 18(2) *Journal of Environmental Assessment Policy and Management* 1601002

Kirchner, Emil and Roberto Dominguez, 'Security Governance in a Comparative Regional Perspective', UNU-CRIS Working Papers W-2013/8 (United Nations University, Brugge, 2013) 1–28

Kirk, Elizabeth A, 'Noncompliance and the Development of Regimes Addressing Marine Pollution from Land-Based Activities' (2008) 39(3) *Ocean Development and International Law* 235–256

Koh, Kheng-Lian, ASEAN *Environmental Law, Policy and Governance Selected Documents* (Volumes I and II) (World Scientific, Singapore, 2013)

Koh, Kheng-Lian, 'ASEAN Environmental Protection in Natural Resources and Sustainable Development: Convergence versus Divergence?' (2007) 4(1) *Macquarie Journal of International and Comparative Environmental Law* 43–73

Koh KL and MS Karim, 'SE Asian Legal Governance', in S Alam, HJH Bhuiyan, TMR Chowdhury and EJ Techera, *Routledge Handbook of International Environmental Law* (Routledge, Abingdon, 2013)

Koh, Kheng-Lian and Nicholas A Robinson, 'Regional Environmental Governance: Examining the Association of Southeast Asian Nations (ASEAN) Model', in Daniel C Esty and Maria H Ivanova (eds), *Global Environmental Governance: Options and Opportunities* (Yale School of Forestry and Environmental Studies, 2002) 101–120

Koh, Kheng-Lian and Nicholas A Robinson, 'Strengthening Sustainable Development in Regional Inter-Governmental Governance: Lessons from the "ASEAN Way"' (2002) 6 *Singapore Journal of International and Comparative Law* 640–682

Koh, Matthew, 'Indicators in the ASEANstats Statistical Regime: A Case-Study on the Need for Accountability, Participation, and Transparency in International Governance by Indicators' (2016) 6(1) *Asian Journal of International Law* 159–191

Koivurova, Timo, 'Could the Espoo Convention become a Global Regime for Environmental Impact Assessment and Strategic Environmental Assessment?', in Robin Warner and Simon Marsden (eds), *Transboundary Environmental Governance: Inland, Coastal and Marine Perspectives* (Ashgate Publishing, Farnham, UK, 2012) 323–342

Koivurova, Timo, 'Transboundary Environmental Impact Assessment in International Law', in Simon Marsden and Timo Koivurova (eds), *Transboundary Environmental Impact Assessment in the European Union: The Espoo Convention and its Kiev Protocol on Strategic Environmental Assessment* (Earthscan, Abingdon, UK and New York, USA, 2011) 15–38.

Koivurova, Timo, Pamela Lesser, Sonia Bickford, Paula Kankaanpää and Marina Nenasheva, *Environmental Impact Assessment in the Arctic* (Edward Elgar, Cheltenham, 2016)

Koivurova, Timo and Tianbao Qin (eds), *Arctic Law and Governance: The Role of China and Finland* (Hart, Oxford, 2016)

Koivurova, Timo, Paula Kankaanpää and Adam Stępień, 'Innovative Environmental Protection: Lessons from the Arctic' (2015) 27 *Journal of Environmental Law* 285–311

Korea Environment Institute, Ministry of Environment, UNECE Espoo Convention, *Workshop on Environmental Impact Assessment in a Transboundary Context in Eastern Asia*, May 2013

Korea Environment Institute, Ministry of the Environment Korea, UNECE Espoo Convention, *Workshop on Transboundary Environmental Impact Assessment in East Asia*, Seoul, South Korea, 13–15 June 2012

Korporaal, Glenda, 'Asian Infrastructure Bank Open', *The Weekend Australian*, 2–3 January 2016

Kotzé, LJ, *Global Environmental Governance: Law and Regulation for the 21st Century* (Edward Elgar, Cheltenham, 2012)

Kotzé, LJ and T Thilo Marauhn (eds), *Transboundary Governance of Biodiversity* (Brill Nijhoff, Leiden and Boston, 2014)

Koyano, Mari, 'Comments', in Jörg Balsiger and Aysun Uyar (eds), *Comparing Regional Environmental Governance in East Asia and Europe: Proceedings* (Research Institute for Humanity and Nature, Kyoto, 2013) 93–96

Kraska, James, 'Coalition Strategy and the Pirates of the Gulf of Aden and the Red Sea' (2009) 28 *Comparative Strategy* 197–216

Kumar, Shashank, 'The Indus Waters Kishenganga Arbitration (Pakistan v. India)' (2013) 17(30) *ASIL Insights*, 13 May 2013

Labarre, Frederic and Maria Tzankova, *Theoretical and Technical Perspectives on Security Sector Governance from the Adriatic to the Caspian Sea* (IOS Press, Amsterdam, 2011)

Laffoley, D and J Langley (eds), *The Bahrain Action Plan for Marine World Heritage: Identifying Priorities for Marine World Heritage and Enhancing the Role of the World Heritage Convention in the IUCN WCPA Marine Global Plan of Action for MPAs in our Oceans and Seas* (IUCN, Gland, 2010)

Lapidoth-Eschelbacher, Ruth, *The Red Sea and the Gulf of Aden* (Martinus Nijhoff, The Hague, 1982)

Larson, Christina, '"Third Pole" Glacier Research Gets a Boost From China' (2011) 334 *Science* 1199

Laruelle, Marlène, 'Russia and Central Asia', in Kavalski, E (ed), *The New Central Asia: The Regional Impact of International Actors* (World Scientific, Singapore, 2010) 149–176

Laurie, Andrew, Jargal Jamsranjav, Onno van den Heuvel and Erdenesaikhan Nyamjav, 'Biodiversity Conservation and the Ecological Limits to Development Options in the Mongolian Altai: Formation of a Strategy and Discussion of Priorities' (2010) 29(3) *Central Asian Survey* 321–343

Lazarus, Kate, Nathan Badenoch, Dao Nga and Bernadette P Resurreccion (eds), *Water Rights and Social Justice in the Mekong Region* (Earthscan, London, 2011)

Lebarre, F and M Tzankova, *Theoretical and Technical Perspectives on Security Sector Governance from the Adriatic to the Caspian Sea* (IOS Press, Amsterdam, 2011)

Lee, Dae-In, Ki-Hyuk Eoma, Gui-Young Kim and Gun-WookBaeck, 'Scoping the Effective Marine Environmental Assessment of Dredging and Ocean Disposal of Coastal Sediments in Korea' (2010) 34 *Marine Policy* 1082–1092

Lee, Janice Ser Huay, David P Bickford, Luis R Carrasco, J Jackson Ewing, Zeehan Jaafar, Lian Pin Koh, Alan Khee Jin Tan and Edward L Webb, 'Toward Clearer Skies: Challenges in Regulating Transboundary Haze in Southeast Asia' (2016) 55(1) *Environmental Science and Policy* 87–95

Lee, J-H, 'Transboundary Pollution in Northeast Asia: An International Environmental Law Perspective' (2013) 35 *University of Hawaii Law Review* 769–785

Lee, Sung-woo, Geun-sub Kim and Eun-woo Kim, *Evaluation Study on the Sea-Land Routes in Northeast Asia* (Greater Tumen Initiative/Korea Maritime Institute, Beijing, 2014)

Lee, Yusin, 'Toward a New International Regime for the Caspian Sea' (2005) 52(3) *Problems of Post-Communism* 37–48

Lenin, Janaki, 'Indigenous Tribe Opposes Hydropower Projects in the Tawang Basin', *The Guardian*, 22 August 2015

Li Hak Yin and Zhengxu Wang, 'China's Advancements in Central Asia: Limitations Toward a Regional Power' (2009) 1(4) *East Asian Policy* 5–13

Li, Jing, 'A Fifth of Glaciers Lost Since the 1950s', *South China Morning Post*, 15 December 2014

Libert, Bo and Annukka Lipponen, 'Challenges and Opportunities for Transboundary Water Cooperation in Central Asia: Findings from UNECE's Regional Assessment and Project Work' (2012) 28(3) *International Journal of Water Resources Development* 565–576

Libert, Bo, Erkin Orolbaev and Yuri Steklov, 'Water and Energy Crisis in Central Asia' (2008) 6(3) *China and Eurasia Forum Quarterly* 9–20

Lidskog, Rolf, 'Extreme Events, Regulatory Style and Regional Environmental Governance', in Jörg Balsiger and Aysun Uyar (eds), *Comparing Regional Environmental Governance in East Asia and Europe: Proceedings* (Research Institute for Humanity and Nature, Kyoto, 2013) 53–63

Lim, Michelle, 'Laws, Institutions and Transboundary Pasture Management in the High Pamir and Pamir-Alai Mountain Ecosystem of Central Asia' (2012) 8(3) *Law, Environment and Development Journal* 43

Lim, Michelle, 'Effective Governance across Central Asian Boundaries: What Lessons does the Central Asian Transboundary Watercourse Management Experience Provide for Mountain Biodiversity Conservation across the Kyrgyz-Tajik Boundary?', in Robin Warner and Simon Marsden (eds), *Transboundary Environmental Governance: Inland,*

Coastal and Marine Perspectives (Ashgate Publishing, Farnham, UK, 2012) 113–140

Lim, Tai-Wei, *Oil and Gas in China: The New Energy Superpower's Relations with its Region* (World Scientific, Singapore, 2010)

Lintner, Stephen, Sherif Arif and Mario Hatziolos, *The Experience of the World Bank in the Legal, Institutional and Financial Aspects of Regional Environmental Programs: Potential Applications of Lessons Learned for the ROPME and PERSGA Programs* (World Bank, Washington DC, 1996)

Litta, H, *Regimes in Southeast Asia: An Analysis of Environmental Cooperation* (Verlag Springer, Wiesbaden, 2012)

Litta, H, 'Regime Theory' in H Litta, *Regimes in Southeast Asia: An Analysis of Environmental Cooperation* (Verlag Springer, Wiesbaden, 2012) 45–74

Liu, Minkang, 'Navigating China's New Silk Road', *South China Morning Post*, 18 January 2016

Liu, Nengye, 'Protection of the Marine Environment from Offshore Oil and Gas Activities', in Rosemary Rayfuse (ed), *Research Handbook on International Marine Environmental Law* (Edward Elgar, Cheltenham, 2015)

Liu, Zhen, 'China Pledges Billions to Mekong River Countries in Bid to Boost Influence and Repair Reputation Amid Tensions in South China Sea', *South China Morning Post*, 24 March 2016

Liu, Zhen, 'China Hopes Mekong Nations Summit will Revive Infrastructure Schemes Amid Criticism of Huge Dam Projects on Upper Reaches of River', *South China Morning Post*, 22 March 2016

Liu, Zhen, 'China Floats Plans for International Maritime Legal Centre Amid Territorial Disputes', *South China Morning Post*, 13 March 2016

Lloyd, SJ, 'Landlocked Central Asia' (1997) 2(1) *Geopolitics* 97

Lo, Chi, 'Why AIIB is Important to China's Economic Health', *Hong Kong Economic Journal*, 22 June 2015

Lo, Sonny, 'Yesterday's Middle Kingdom, China Today Seeks to be the Heartland of Asia', *South China Morning Post*, 8 December 2014

Loure, Flavia Rocha and Alistair Rieu-Clarke (eds), *The UN Watercourses Convention in Force: Strengthening International Law for Transboundary Water Management* (Routledge, Abingdon, UK, 2013)

Lovett, Richard A, 'Man-Made Earthquakes Cause Seismic Rumblings', *Cosmos Magazine*, 24 March 2014

Lu, Z, DG Streets, Q Zhang and S Wang, 'A Novel Back-Trajectory Analysis of the Origin of Black Carbon Transported to the Himalayas and Tibetan Plateau during 1996–2010' (2012) 39 *Geophysical Research Letters* LO1809

Lyons, Youna 'The New Offshore Oil and Gas Installation Abandonment Wave and the International Rules on Removal and Dumping' (2014) 29 *International Journal of Marine and Coastal Law* 498

Mackay, Joseph, 'Running Dry: International Law and the Management of the Aral Sea Depletion' (2009) 28(1) *Central Asian Survey* 17–27

Magee, Darrin, 'China Fails to Build Trust with Mekong Neighbours', in China Dialogue, *Forging a New Course for the Mekong* 8–10

Magsig, Bjørn-Oliver, *International Water Law and the Quest for Common Security* (Earthscan/Routledge, Abingdon and New York, 2015)

Mahmoudi, Said, 'Legal Protection of the Persian Gulf's Marine Environment' (1997) 21(1) *Marine Policy* 53–62

Majumder, Sanjoy, 'Why India is Concerned about Nepal's Constitution', *The Guardian*, 22 September 2015

Malik, H (ed), *Central Asia: Its Strategic Importance and Future Prospects* (St Martin's Press, New York, 1994)

Marketos, TN, *China's Energy Geopolitics: The Shanghai Cooperation Organization and Central Asia* (Routledge, Abingdon and New York, 2009)

Marsden, Simon, 'From the High North to the Roof of the World: Arctic Precedents for Third Pole Governance' (2016) 8 *The Yearbook of Polar Law* 56–75

Marsden, Simon, 'Environmental Assessment of Cross-Border Development: China and the Third Pole' (2016) 18 *Journal of Environmental Assessment Policy and Management* 1650009-1-1650009-21

Marsden, Simon, 'The Helsinki Water Convention: Implementation and Compliance in Asia' (2015) 20(2) *Nordic Environmental Law Journal* 119–129

Marsden, Simon, 'The World Heritage Convention: Compliance, Public Participation and the Rights of Indigenous People' (2015) 32(6) *Environmental and Planning Law Journal* 534–547

Marsden, Simon, 'Asian Precedents for Effective Biodiversity Conservation in the Third Pole', paper presented to *International Symposium: Biodiversity, Sustainable Development and the Law*, Centre for International Sustainable Development Law/Lauterpacht Centre for International Law, St John's College, University of Cambridge, 2015

Marsden, Simon, 'The World Heritage Convention in the Arctic and Indigenous People: Time to Reform?' (2014) 6 *The Yearbook of Polar Law* 226–249

Marsden, Simon 'Protecting Heritage on Australia's Coasts: A Role for Strategic Environmental Assessment?' (2013) 15(3) *Journal of Environmental Assessment Policy and Management* 1350014-1-1350014-21

Marsden, Simon, 'Regulatory Reform of Australia's Offshore Oil and Gas Sector after the Montara Commission of Inquiry: What about

Transboundary Environmental Impact Assessment?' (2013) 15(1) *Flinders Law Journal* 41–53

Marsden, Simon, 'Cooperative Environmental Planning in China's Pearl River Delta Region', in C Andressen, AR Mubarak and X Wang (eds), *Sustainable Development in China* (Routledge, Abingdon, UK, 2013) 152–163

Marsden, Simon, 'Developing Agreements for Transboundary Environmental Impact Assessment and Strategic Environmental Assessment in Asia', in Robin Warner and Simon Marsden (eds), *Transboundary Environmental Governance: Inland, Coastal and Marine Perspectives* (Ashgate, Farnham, UK, 2012) 141–164

Marsden, Simon, 'Public Participation in Transboundary Environmental Impact Assessment – Closing the Gap between International and Public Law?', in Brad Jessup and Kim Rubenstein (eds), *Environmental Discourses in Public and International Law* (Cambridge University Press, Cambridge, 2012) 238–259

Marsden, Simon, 'Invoking Direct Application and Effect of International Treaties by the European Court of Justice: Implications for International Environmental Law in the European Union' (2011) 60(3) *International and Comparative Law Quarterly* 737–757

Marsden, Simon, 'Assessment of Transboundary Environmental Effects in the Pearl River Delta Region: Is There a Role for Strategic Environmental Assessment?' (2011) 31(6) *Environmental Impact Assessment Review* 593–601

Marsden, Simon, 'Introducing Strategic Environmental Assessment to the Madrid Protocol: Lessons from International Experience' (2011) 1(1) *The Polar Journal* 33–47

Marsden, Simon, 'Developing Approaches to Transboundary Environmental Impact Assessment in China: Cooperation through the Greater Tumen Initiative and in the Pearl River Delta Region' (2010) 9(2) *Chinese Journal of International Law* 393–414

Marsden, Simon, 'Enforcing Non-Discrimination in Transboundary Environmental Impact Assessment: Advantages for EU Citizens from the Transposition of the Espoo and Aarhus Conventions?' (2009) 6 *Journal for European Environmental and Planning Law* 437

Marsden, Simon, *Strategic Environmental Assessment in International and European Law* (Earthscan, Abingdon and New York, 2008)

Marsden, Simon, 'Importance of Context in Measuring the Effectiveness of Strategic Environmental Assessment' (1998) 16(4) *Impact Assessment and Project Appraisal* 255–266

Marsden, Simon and Elizabeth Brandon, *Transboundary Environmental Governance in Asia: Practice and Prospects with the UNECE Environmental Agreements* (Edward Elgar, Cheltenham, 2015)

Marsden, Simon and Timo Koivurova (eds) *Transboundary Environmental Impact Assessment in the European Union: The Espoo Convention and its Kiev Protocol on Strategic Environmental Assessment* (Earthscan/Routledge, Abingdon and New York, 2011)

Maruma Mrema, Elizabeth, 'Regional Seas Programme: The Role Played by UNEP in its Development and Governance', in David Attard, Malgosia Fitzmaurice, Norman Martinez and Riyaz Hamza (eds), *The IMLI Manual on International Maritime Law. Vol. III, Marine Environmental Law and International Maritime Security Law* (Oxford University Press, Oxford, 2016)

Mauerhofer, Volker, Edwald Galle and Marco Onida, 'The Alps Convention', in CJ Bastmeijer (ed), *Wilderness Protection in Europe: The Role of International, European and National Law* (Cambridge University Press, Cambridge, 2016) 199–221

Maverick, Tim, 'A Thai Canal to be Part of China's Silk Road', *Wall Street Daily*, 27 January 2016

McKinney, Daene C, 'Cooperative Management of Transboundary Water Resources in Central Asia', in A Dinar, S Dinar, S McCaffrey and D Mckinney, *Bridges over Waters: Understanding Transboundary Water Conflict, Negotiation and Cooperation* (World Scientific, Singapore, 2013) 187–219

Mehdiyoun, Kamyar, 'International Law and the Dispute over Ownership of Oil and Gas Resources in the Caspian Sea' (2000) 94(1) *The American Journal of International Law* 179–189

Mekong River Commission for Sustainable Development, *Guidelines for Implementation of the Procedures for Notification, Prior Consultation and Agreement*, 21 August 2005

Mekong River Commission for Sustainable Development, *Procedures for Notification, Prior Consultation and Agreement*, Phnom Penh, 13 November 2003

Mekong River Commission for Sustainable Development, *1995 Mekong Agreement and Procedures*, undated

Mekong River Commission for Sustainable Development, *Public Participation in the Context of the MRC*, undated

Mekong River Commission, 'China Commits to Continue and Enhance Cooperation with the MRC', *News*, 7 November 2014

Meng, Angela, 'Central Asian Nations to Conduct Terrorism Drills with PLA', *South China Morning Post*, 14 August 2014

Meng, Si, 'The Fate of People Displaced by the Mekong Dams', in China Dialogue, *Forging a New Course for the Mekong* 25–27

Miles, Edward L, Steiner Andresen, Elaine M Carlin, Jon Birger Skjaerseth, Arild Underdal and Jorgen Wetterstad, *Environmental Regime*

Effectiveness: Confronting Theory with Evidence (MIT Press, Cambridge, 2002)

Milev, EL, A Underdal, S Andresen, J Wetterstad, JB Skjaerseth and EM Carlin, *Environmental Regime Effectiveness: Confronting Theory with Evidence* (MIT Press, Cambridge, 2002)

Mirza, Monirul Qader, 'Climate Change, Flooding in South Asia and Implications' (2011) 11 *Regional Environmental Change* S95–S107

Molden, D and E Sharma, 'ICIMOD's Strategy for Delivering High-quality Research and Achieving Impact for Sustainable Mountain Development' (2013) 33(2) *Mountain Research and Development* 179–183

Molle, François, Tira Foran and Mira Kakonen (eds), *Contested Landscapes in the Mekong Region: Hydropower, Livelihoods and Governance* (Earthscan, Abingdon and New York, 2009)

Monbiot, George, 'Why the EU's Increasing Failure to Protect Nature Means I May Vote No', *The Guardian*, 10 July 2015

Moore, Malcolm, 'Chinese Earthquake May Have Been Man-Made, Say Scientists', *The Telegraph*, 2 February 2009

Moorman, Jesse L and Zhang Ge, 'Promoting and Strengthening Public Participation in China's Environmental Impact Assessment Process: Comparing China's EIA Law and U.S. NEPA' (2007) 8 *Vermont Journal of Environmental Law* 281–335

Morton, Katherine, 'Climate Change and Security at the Third Pole' (2011) 53(1) *Survival: Global Politics and Strategy* 121–132

Mourant, Alex, Douglas Emeott, Jagabanta Ningthoujam and Jasmin Yu, *Ensuring Sustainability in the Asian Infrastructure Investment Bank and the New Development Bank, A SAIS International Development Practicum Report* (John Hopkins School of Advanced International Studies, 2015)

Mousavi, Seyyed Rasoul, 'The Future of the Caspian Sea after Tehran Summit' (2008–2009) XXI(1–2) *The Iranian Journal of International Affairs* 27–42

Moynihan, Ruby and Bjørn-Oliver Magsig, 'The Rising Role of Regional Approaches in International Water Law: Lessons from the UNECE Water Regime and Himalayan Asia for Strengthening Transboundary Water Cooperation' (2014) 23(1) *Review of European, Comparative and International Environmental Law* 43–58

Mrema, Elizabeth Maruma, 'Regional Seas Programme: The Role Played by UNEP in its Development and Governance', in David Attard, Malgosia Fitzmaurice, Norman Martinez and Riyaz Hamza (eds), *The IMLI Manual on International Maritime Law. Vol. III, Marine Environmental Law and International Maritime Security Law* (Oxford University Press, Oxford, 2016)

Muni, Sukh Deo and Rajshree Jetly, 'SAARC: The Changing Dimensions', UNU-CRIS Working Papers W-2008/8 (United Nations University, 2008) 1–26

Mushkat, Roda, *International Environmental Law and Asian Values: Legal Norms and Cultural Influences* (UBC Press, Vancouver, 2004)

Mushkat, Roda, 'Creating Regional Environmental Governance Regimes: Implications of Southeast Asian Responses to Transboundary Haze Pollution' (2013) 4(1) *Washington and Lee Journal of Energy, Climate and the Environment* 103–160

Myint, Tun, *Governing International Rivers: Polycentric Politics in the Mekong and the Rhine* (Edward Elgar, Cheltenham, 2012)

Nadim, Farhad, Amvrossios C Bagtzoglou and Jamshid Iranmahboob, 'Coastal Management in the Persian Gulf Region within the Framework of the ROPME Programme of Action' (2008) 51 *Ocean and Coastal Management* 556–565

Nam, Sangmin, 'Ecosystem Governance in a Cross-border Area: Building a Tumen River Trans-boundary Biosphere Reserve' (2005) 7 *China Environment Series* 83–88

Nam, Sangmin and Heejoo Lee, 'Reverberating beyond the Region in Addressing Air Pollution in Northeast Asia', in Jörg Balsiger and Aysun Uyar (eds), *Comparing Regional Environmental Governance in East Asia and Europe: Proceedings* (Research Institute for Humanity and Nature, Kyoto, 2013) 75–89

Naser, Humood, John Bythell and Jeremy Thomason, 'Ecological Assessment: An Initial Evaluation of the Ecological Input in Environmental Assessment Reports in Bahrain' (2008) 26(3) *Impact Assessment and Project Appraisal* 201–208

National Environment Agency, 'Factsheet on Transboundary Haze Pollution' (Government of Singapore, 2014)

Nazari, Mehrdad, 'The Transboundary EIA Convention in the Context of Private Sector Operations Co-financed by an International Financial Institution: Two Case Studies from Azerbaijan and Turkmenistan' (2003) 23 *Environmental Impact Assessment Review* 441

Newell, Peter, 'The Political Economy of Global Environmental Governance' (2008) 34(3) *Review of International Studies* 507–529

Newton, Scott, *The Constitutional Systems of the Independent Central Asian States: A Contextual Analysis* (Hart Publishing, Oxford, 2016)

Ng, Eric, 'China's Nuclear Plant Makers Seek New Markets Along the Ancient Silk Road into Asia, Europe, Africa and Middle East', *South China Morning Post*, 3 April 2016

Ng, Teddy, 'Beijing "May Need Rethink" as Vietnam Moves Closer to US', *South China Morning Post*, 30 May 2016

Nguitragool, P, *Environmental Cooperation in Southeast Asia: ASEAN's Regime for Transboundary Haze Pollution* (Routledge, London, 2011)

Niazi, Tarique, 'The Ecology of Strategic Interests: China's Quest for Energy Security from the Indian Ocean and the South China Sea to the Caspian Sea Basin' (2006) 4(4) *China and Eurasian Forum Quarterly* 97–116

Nouri, J, AR Karbassi and S Mirkia, 'Environmental Management of Coastal Regions in the Caspian Sea' (2008) 5(1) *International Journal of Environmental Science and Technology* 43–52

Nurhidayah, Laely, Zada Lipman and Shawkat Alam, 'Regional Environmental Governance: An Evaluation of the ASEAN Legal Framework for Addressing Transboundary Haze Pollution' (2014) 15(1) *Australian Journal of Asian Law* 1–17

Ogden, Christopher, *China and India: Asia's Emergent Great Powers* (Polity, Cambridge, in press)

Ogden, Christopher (eds), *New South Asian Security: Six Core Relations Underpinning Regional Security* (Orient BlackSwan, Telangana, 2016)

Omar, Samira AS, Ernest Briskey, Raafat Misak and Adel ASO Asem, 'The Gulf War Impact on the Terrestrial Environment of Kuwait: An Overview', in Jay E Austin and Carl E Bruch (eds), *The Environmental Consequences of War: Legal, Economic, and Scientific Perspectives* (Cambridge University Press, Cambridge, 2007) 316–337

Oral, Nilufer, 'Forty Years of the UNEP Regional Seas Programme: From Past to Future', in Rosemary Rayfuse (ed), *Research Handbook on International Marine Environmental Law* (Edward Elgar, Cheltenham, 2015)

Ostrowski, Wojciech, *Politics and Oil in Kazakhstan* (Routledge, London, 2009)

Pagnamenta, Robin, 'Rise of Saudi Crown Prince', *The Times*, 23 April 2016

Pak, Ali and Manuchehr Farajzadeh, 'Iran's Integrated Coastal Management Plan: Persian Gulf, Oman Sea, and Southern Caspian Sea Coastlines' (2007) 50 *Ocean and Coastal Management* 754–773

Park Lee, A and S Matsuoka, 'Effectiveness of International Institutions: How to Build an Analytical Framework?', paper presented to the *Conference on the Human Dimensions of Global Environmental Change*, Amsterdam, December 2–4, 2009

Pomfret, Richard, *Regionalism in East Asia: Why has it Flourished since 2000 and How Far Will it Go?* (World Scientific, Singapore, 2011)

Pomfret, Richard, 'The Tumen River Area Development Programme' Winter (1997–1998) *IBRU Boundary and Security Bulletin* 80–88

Pomfret, Richard, 'Central Asia since the Dissolution of the Soviet Union: Economic Reforms and their Impact on State-Society Relations', in MP Amineh (ed), *The Greater Middle East in Global Politics: Social Science Perspectives on the Changing Geography of the World Politics* (Brill, Leiden, 2007) 301–332

Ponte, Elena, 'Preliminary Analysis of Water Security in the Mekong River Basin', UNU-CRIS Working Papers W-2012/12 (United Nations University, Brugge, 2012) 1–25

Poole, Thom, 'Iran and Saudi Arabia's Great Rivalry Explained', *BBC News*, 4 January 2016

Porkareh, Mohammad Hossein, Seyed Masood Monavari, Yousef Molaei and Masood Akhavan Fard, 'Comparison of Regional Organisations for Protection of the Marine Environment (ROPME) and the Convention of Caspian Sea Provisions' (2013) 4(7) *Annals of Biological Research* 27–34

Prashant, Shishir, 'Jammu-Pak Border Might Have Oil and Gas, Says ONGC', *Business Standard*, 7 February 2014

Price, ARG, 'Impact of the 1991 Gulf War on the Coastal Environment and Ecosystems: Current Status and Future Prospects' (1998) 24 (1–2) *Environment International* 91–96

Price, Gareth, Rafay Alam, Sarah Hasan, Fahd Humayun, Mohammad Humayun Kabir, Chandni Singh Karki, Sonali Mittra, Tasmia Saad, Mehr Saleem, Samir Saleem, Pragya Ratna Shakya, Christopher Snow and Shayasta Tuladhar, *Attitudes to Water in South Asia* (Chatham House, London, 2014)

Putin, Vladimir, 'Family of Nations: The Key to Progress and Development in the Asia-Pacific Region Lies in Ever-Closer Cooperation', *South China Morning Post*, 17 November 2015

Putz, Catherine, 'Uzbekistan Still Hates the Rogun Dam Project', *The Diplomat*, 4 August 2015

Qiu, Jane, 'China: The Third Pole' (2008) 454 *Nature* 393–396

Qiu, Jane, 'Glaciologists to Target Third Pole' (2012) 484 *Nature* 19

Qoraboyev, Ikboljon, 'Around the Names of Regions: The Case of Central Asia', UNU-CRIS Working Papers, W-2010/5 (United Nations University, 2010) 1–34

Radtke, Kurt, 'China and the Greater Middle East: Globalization No Longer Equals Westernization', in MP Amineh (ed), *The Greater Middle East in Global Politics: Social Science Perspectives on the Changing Geography of the World Politics* (Brill, Leiden, 2007) 389–416

Rahimov, Mirzohid, 'From Soviet Republics to Independent Countries: Challenges of Transition in Central Asia', in MP Amineh (ed), *The*

Greater Middle East in Global Politics: Social Science Perspectives on the Changing Geography of the World Politics (Brill, Leiden, 2007) 291–312

Razzaque, Jona, *Environmental Governance in Europe and Asia: A Comparative Study of Institutional and Legislative Frameworks* (Routledge, Abingdon, 2013)

Razzaque, Jona, 'Public Participation in Water Governance', in Joseph W Dellapenna and Joyetta Gupta (eds), *The Evolution of the Law and Politics of Water* (Springer, 2009) 353–372

Reel, Yesim, 'Energy', in EAH Gencer and C Gerni (eds), *Central Asia Economies in Transition* (Cambridge Scholars, Newcastle, 2012)

Regional Mountain Centre for Central Asia, *Strategy of the Regional Mountain Centre of Central Asia*, Bishkek, the Kyrgyz Republic, 2009

Regional Organization for the Protection of the Marine Environment, *Guidelines for Compensation Following Damage to Coral Reefs by Ship or Boat Grounding*, Technical Series Number 15, February 2009

Regional Organization for the Conservation of the Environment of the Red Sea and Gulf of Aden (PERSGA), *EIA and EMS Guidelines for Fishery/Aquaculture Projects in the Red Sea and Gulf of Aden*, PERSGA Technical Series No. 9, May 2004

Regional Organization for the Conservation of the Environment of the Red Sea and Gulf of Aden, *Regional Action Plan for the Conservation of Coral Reefs in the Red Sea and Gulf of Aden*, PERSGA Technical Report Series No. 3, PERSGA Jeddah 2003

Regional Organization for the Conservation of the Environment of the Red Sea and Gulf of Aden, *The Red Sea and Gulf of Aden Regional Network of Marine Protected Areas – Regional Master Plan*, PERSGA Technical Report Series No. 1, PERSGA Jeddah 2002

Regional Organization for the Conservation of the Environment of the Red Sea and Gulf of Aden, *Final Act of the Jeddah Plenipotentiary Regional Conference on the Conservation of the Marine Environment and Coastal Areas in the Red Sea and Gulf of Aden* 1982

Regional Organization for the Protection of the Marine Environment, *Guidelines on Requirements for the Conduct of Environmental Impact Surveys and the Production of Environmental Impact Statements*, MEMAC, undated

Regional Organization for the Protection of the Marine Environment, *Guidelines on the Voluntary Regulation of the Disposal of Drill Cuttings on the Sea Bed*, MEMAC, undated

Regional Organization for the Protection of the Marine Environment, *Guidelines on the Use and Storage of Chemicals in Offshore Operations*, MEMAC, undated

Regional Organization for the Protection of the Marine Environment, *Guidelines on the Conduct of Seismic Operations*, MEMAC, undated

Renaud, Fabrice G and Claudia Kuenzer (eds), *The Mekong Delta System: Interdisciplinary Analyses of a River Delta* (Springer, Dordrecht, 2012)

Reuters, 'Japan's Mekong Aid May Counter China', *South China Morning Post*, 3 May 2016

Reuters, 'China's "New Silk Road" Facing Difficulties: Former Official', *South China Morning Post*, 23 April 2016

Reuters, 'China to Build Second Railway Line into Tibet', *The Guardian*, 5 March 2016

Reuters, 'ASEAN Expresses Serious Concern Over South China Sea Developments', *EJ Insight*, 29 February 2016

Reuters, 'China and India Call on Rich Countries to Step up Climate Change Efforts', *The Guardian*, 15 May 2015

Rieu-Clarke, Alistair, 'Notification and Consultation Procedures under the Mekong Agreement: Insights from the Xayaburi Controversy' (2015) 5 *Asian Journal of International Law* 143–175

Roberts, Adam, 'The Arab Spring: Why Did Things Go So Badly Wrong?', *The Guardian*, 15 January 2016

Ronen, Yael, *The Iran Nuclear Issue* (Hart, Oxford, 2010)

Rothwell, Donald R and Tim Stephens, *The International Law of the Sea* (2nd edition, Hart, Oxford, 2016)

Rowe, William C, 'Agrarian Adaptations in Tajikistan: Land Reform, Water and Law' (2010) 29(2) *Central Asian Survey* 189–204

Ruan, Victoria, 'Silk Road Plan Risks Low Business Return', *South China Morning Post*, 5 January 2015

Rudaz, Gilles, 'The Great Mountain Game: Regional Governance in Central Asia', in Jörg Balsiger and Aysun Uyar (eds), *Comparing Regional Environmental Governance in East Asia and Europe: Proceedings* (Research Institute for Humanity and Nature, Kyoto, 2013) 43–50

Saddique, Haroon and Jason Burke, 'China May Build Rail Tunnel Under Mount Everest, State Media Reports', *The Guardian*, 9 April 2015

Sadiq, Muhammad and John Charles McCain, *The Gulf War Aftermath: An Environmental Tragedy* (Springer, 1993)

Salman, Ali, 'Pakistan-China Economic Corridor: A Cost Benefit Analysis', *The Express Tribune*, 3 May 2015

Salman, MA Salman, *The World Bank Policy for Projects on International Waterways: An Historical and Legal Analysis* (The World Bank, Washington DC, 2009)

Sarfraz, Hamid, 'Revisiting the 1960 Indus Waters Treaty' (2013) 38(2) *Water International* 204–216

Saul, Ben, 'China, Sovereignty, Natural Resources and International Law', in Kate Barclay and Graeme Smith (eds), *East Asia's Demand for Energy, Minerals and Food: The International Politics of Resources* (Routledge, London, 2015)

Saul, Ben and Tim Stephens (eds), *Antarctica in International Law* (Hart Publishing, Oxford, 2015)

Save the Mekong Coalition, 'Request for Clarification Regarding the First Regional Consultation on the Don Sahong Hydropower Project', Letter to Mr. Hans Guttman, CEO of the Mekong River Commission, 11 November 2014

Schrage, Wiek, Keith Bull and Albena Karadjova, 'Environmental Legal Instruments in the UNECE Region' (2008) 18(1) *Yearbook of International Environmental Law* 3–31

Schreve, Leonie, 'The Equator Principles: A Voluntary Approach by Bankers', in Kees Bastmeijer and Timo Koivurova (eds), *Theory and Practice of Transboundary Environmental Impact Assessment* (Martinus Nijhoff, Leiden and Boston, 2008) 327–344

See-Won Byun, 'Provincial Impacts of Multilateral Cooperation: The Greater Tumen Initiative and Environmental Protection in Jilin' (2008/2009) 10 *China Environment Series* 163

Sheppard, Charles, Mohsen Al-Husiani, F Al-Jamali, Faiza Al-Yamani, Rob Baldwin, James Bishop, Francesca Benzoni, Eric Dutrieux, Nicholas K Dulvy, Subba Rao V Durvasula, David A Jones, Ron Loughland, David Medio, M Nithyanandan, Graham M Pilling, Igor Polikarpov, Andrew RG Price, Sam Purkis, Bernhard Riegl, Maria Saburova, Kaveh Samimi Namin, Oliver Taylor, Simon Wilson and Khadija Zainal, 'The Gulf: A Young Sea in Decline' (2010) 60 *Marine Pollution Bulletin* 13–38

Shi, Jiangtao, 'South China Sea Tensions to Dominate Sino-Asean Talks in Singapore', *South China Morning Post*, 27 April 2016

Shiraiwa, Takayuki, 'Amur-Okhotsk Consortium as an Epistemic Community to Conserve the Transboundary Land-Ocean Ecosystem in Northeast Asia and the Russian Far East', in Jörg Balsiger and Aysun Uyar (eds), *Comparing Regional Environmental Governance in East Asia and Europe: Proceedings* (Research Institute for Humanity and Nature, Kyoto, 2013) 29–32

Sitaraman, Srini, *State Participation in International Treaty Regimes* (Ashgate, Farnham, UK, 2009)

Slocum, Nikki and Luk Van Langenhove, 'The Meaning of Regional Integration: Introducing Positioning Theory in Regional Integration

Studies', UNU-CRIS e-Working Papers W-2003/5 (United Nations University, 2003) 1–49

Smith, Elizabeth, 'Implementing the Espoo Convention: An International Financial Institution Perspective', in Kees Bastmeijer and Timo Koivurova (eds), *Theory and Practice of Transboundary Environmental Impact Assessment* (Martinus Nijhoff, Leiden and Boston, 2008) 313–326

Sneddon, Chris and Coleen Fox, 'Rethinking Transboundary Waters: A Critical Hydropolitics of the Mekong Basin' (2006) 25 *Political Geography* 181–202

Söderbaum, F and J Granit, *The Political Economy of Regionalism: The Relevance for International Waters and the Global Environment Facility* (Global Environment Facility, Washington DC, 2014)

Sornarajah, M and Jiangyu Wang, *China, India and the International Economic Order* (Cambridge University Press, Cambridge, 2010)

Steiner, Achim, 'Foreword' in *South Asia Environment Outlook 2009* (UNEP/SAARC, 2009)

Stephens, Tim, 'Disasters, International Environmental Law and the Anthropocene', in Susan Breau and Katja LH Samuel (eds), *Research Handbook on Disasters and International Law* (Edward Elgar, Cheltenham, 2016)

Stolberg, F, O Borysova, I Mitrofanov, V Barannik and P Eghtesadi, *Global International Waters Assessment – Caspian Sea*, GIWA Regional Assessment 23 (University of Kalmar/United Nations Environment Programme, 2006)

Sukaroek, Worawan, 'The MRC Policy Framework on Transboundary EIA: A Better Cooperation will be Enhanced in the Lower Mekong Basin', paper presented to the Conference, *Role of Water Sciences in Transboundary River Basin Management*, Thailand, 2005

Swain, A, 'South Asia, its Environment and Regional Institutions', in L Elliott and S Breslin (eds), *Comparative Environmental Regionalism* (Routledge, Abingdon, 2011) 76–91

Takahashi, Wakana, 'Environmental Cooperation in South Asia', in K Kato and W Takahashi, *Regional/Subregional Environmental Cooperation in Asia* (IGES, 2001) 51–63

Takahashi, Wakana, 'Environmental Cooperation in Southeast Asia (ASEAN)', in Kazu Kato and Wakana Takahashi (eds), *Regional/Subregional Environmental Cooperation in Asia* (Institute for Global Environmental Strategies, 2001) 31–50

Talley, Ian, 'Obama: We're All for the Asian Infrastructure Investment Bank', *The Wall Street Journal*, 28 April 2015

Talmon, Jia, *The South China Sea Arbitration* (Hart, Oxford, 2014)

Tam, Yin-Ki, I-Hsun Ni, Cynthia Yau, Man-Yee Yan, Wai-Shan Chan, Sze-Man Chan and Hsueh-Jung Lu, 'Tracking the Changes of a Fish Community Following a Megascale Reclamation and Ensuing Mitigation Measures' (2013) 70(6) *ICES Journal of Marine Science* 1206–1219

Tang, Shui-Yan, Ching-Ping Tang and Carlos Wing-Hung Lo, 'Public Participation and Environmental Impact Assessment in Mainland China and Taiwan: Political Foundations of Environmental Management' (2005) 41 *Journal of Development Studies* 1–32

Tang Tao, Zhu Tan and Xu He, 'Integrating Environment into Land Use Planning through Strategic Environmental Assessment in China: Towards Legal Frameworks and Operational Procedures' (2007) 27 *Environmental Impact Assessment Review* 243–265

Tanzi, Attila, 'Regional Contributions to International Water Cooperation: The UNECE Contribution', in Laurence Boisson de Chazournes, Christina Leb and Mara Tignino (eds), *International Law and Freshwater: The Multiple Challenges* (Edward Elgar, Cheltenham, 2013) 155–178

Tanzi, Attila, Owen McIntyre, Alexandros Kolliopoulos, Alistair Rieu-Clarke and Remy Kinna (eds), *UNECE Convention on the Protection and Use of Transboundary Watercourses and International Lakes: Its Contribution to International Water Cooperation* (Brill Nijhoff, Leiden and Boston, 2015)

Tavares, Rodrigo, 'The State of the Art of Regionalism: The Past, Present and Future of a Discipline', UNU-CRIS Working Papers W-2004/10 (United Nations University, 2004) 1–29

Taylor, Adam, 'Saudi Arabia Wants You to Go on Holiday There. Here's Why', *The Independent*, 27 April 2016

Taylor, Peter Mark, Patricia Charlebois, Violeta Velikova and Frits Schlingemann, 'Oil Spill Preparedness and Co-operation in the Caspian Sea and Black Sea: Regional Successes and Lessons Learned', paper presented to the *International Oil Spill Conference* (2011)

Trachtman, Joel P, 'Persistent Objectors, Cooperation, and the Utility of Customary International Law' (2010) 21 *Duke Journal of Comparative and International Law* 221–233

Tremlett, Giles, 'Catalonia's Separatist Government will Struggle to Break Stalemate with Spain', *The Guardian*, 11 January 2016

Tripathi, Dhananjay, 'Is Economic Growth the Required "Catalyst" for South Asian Integration?', UNU-CRIS Working Papers W-2013/12 (United Nations University, Brugge, 2013) 1–23

Tsering Topgyal, 'China and India's Border Dispute Rises to Dangerous New Heights', *The Conversation*, 26 September 2014

Tstsumi, R and K Robinson, 'Environmental Impact Assessment and the Framework Convention for the Protection of the Marine Environment of the Caspian Sea', in Kees Bastmeijer and Timo Koivurova, *Theory and Practice of Transboundary Environmental Impact Assessment* (Brill Nijhoff, Leiden and Boston, 2007)

Tung, Chee-hwa, 'Right Side of History', *South China Morning Post*, 26 May 2016

Turner, Jennifer and Kenji Otsuka, *Reaching Across the Water: International Cooperation Promoting Sustainable River Basin Governance in China* (Woodrow Wilson Center for Scholars, 2006)

Underdal, Arild, 'One Question, Two Answers', in EL Miles, A Underdal, S Andresen, J Wettestad, JB Skjærseth and EM Carlin, *Environmental Regime Effectiveness: Confronting Theory with Evidence* (MIT Press, Cambridge, 2002) 3–45

United Nations Development Programme and Korean National Commission for UNESCO, *Project Final Report, Feasibility Study on the Establishment of the Lower Tumen River Area Trans-boundary Biosphere Reserve* (2004)

United Nations Economic Commission for Europe, UNECE Espoo Convention on Environmental Impact Assessment Becomes a Global Instrument, Press Release, 27 August 2014

United Nations Economic Commission for Europe, Chair's summary of the meeting 'Further development of regional cooperation among the countries of Central Asia in the management of information on the environment and water resources', Almaty, 9–10 June 2014

United Nations Economic Commission for Europe, Espoo Convention, Meeting of the Parties Sixth session, Geneva, 2–5 June 2014, Item 6 of the provisional agenda, Seminar on the globalization of the Convention and the Protocol and the role of international financial institutions, Information on the seminar on the globalization of the Convention and the Protocol and the role of international financial institutions, ECE/MP.EIA/2014/INF.8, 28 May

United Nations Economic Commission for Europe, Espoo Convention, Meeting of the Parties, Sixth session, Geneva, 2–5 June 2014, Item 9 of the provisional agenda, Panel discussion on the application of the Convention and Protocol to energy-related issues, Informal summary of the panel discussion by the moderator

United Nations Economic Commission for Europe, Espoo Convention, Meeting of the Parties, Sixth session, Geneva, 2–5 June 2014, Items 3(d) and 12 of the provisional agenda, Outstanding issues: draft declaration, Adoption of the declaration, Draft declaration proposal by the Bureau, ECE/MP.EIA/2014/L.4–ECE/MP.EIA/SEA/2014/L.4, Part B, 19 March

United Nations Economic Commission for Europe, Espoo Convention, Meeting of the Parties, Sixth session, Geneva, 2–5 June 2014, Report of the Meeting of the Parties, addendum, decisions and the declaration adopted jointly by the Meeting of the Parties (Convention and Protocol), ECE/MP.EIA/20/Add.3–ECE/MP.EIA/SEA/4/Add.3, 15 July 2014, 31–32

United Nations Economic Commission for Europe, Espoo Convention, Meeting of the Parties Sixth session, Geneva, 2–5 June 2014, Report of the Meeting of the Parties, ECE/MP.EIA/20–ECE/MP.EIA/SEA/4, 15 July 2014

United Nations Economic Commission for Europe, Convention on the Protection and Use of Transboundary Watercourses and Lakes, *The Global Opening of the 1992 UNECE Water Convention* (UNECE, 2013)

United Nations Economic Commission for Europe, Convention on the Protection and Use of Transboundary Watercourses and International Lakes, *Second Assessment of Transboundary Rivers, Lakes and Groundwaters* (United Nations, 2011)

United Nations Economic Commission for Europe, UNECE Welcomes the Russian Federation's Intention to Ratify the Aarhus and Espoo Environmental Conventions, *UNECE Weekly*, 432, 27 June–1 July 2011

United Nations Economic Commission for Europe/Attila Tanzi, *The Economic Commission for Europe Water Convention and the United Nations Watercourses Convention: An Analysis of their Harmonized Contribution to International Water Law* (United Nations, 2015)

United Nations Economic and Social Commission for Asia and the Pacific, NEASPEC *Workshop on Transboundary Air Pollution in North-East Asia, Report of the Meeting*, 10–11 November 2011, Incheon

United Nations Economic and Social Council, *Report of the Working Group on EIA on its 12th Meeting*, Geneva, 11–13 May 2009

United Nations Environment Programme, *The Regional Perspective* (2003)

United Nations Environment Programme, United Nations Economic Commission for Europe, United Nations Development Programme, North Atlantic Treaty Organization (NATO), Regional Environment Centre (REC), Organization for Security and Co-operation in Europe (OSCE), *Environment and Security in Europe, Transforming Risks into Cooperation: The Case of the Eastern Caspian Region* (2008)

United Nations Environment Programme and Caspian Environment Programme, *Guidelines on Environmental Impact Assessment in a Transboundary Context in the Caspian Sea Region: Step by Step Procedures*

(United Nations Environment Programme and Caspian Environment Programme, 2003)
Van den Broek, Marcel, *Transport Corridors in the Greater Tumen Region: Financing Infrastructure Development: Study Report* (Deutsche Gesellschaft für Internationale Zusammenarbeit (GIZ) GmbH/Greater Tumen Initiative, Beijing, 2014)
Van Langenhove, Luk, 'Theorising Regionhood', UNU-CRIS Working Papers, W-2003/1 (United Nations University, Brugge, 2003)
Varkkey, Helena, *The Haze Problem in Southeast Asia: Palm Oil and Patronage* (Routledge, Abingdon, 2016)
Vidal, John, 'Most Glaciers in Mount Everest Area will Disappear with Climate Change – Study', *The Guardian*, 27 May 2015
Vidal, John and K Dasgupta, 'Himalayas at Risk in Hydro Dam Water Grab', *New Zealand Herald News*, 13 August 2013
Vinogradov, Sergei, 'The Impact of the Deepwater Horizon: The Evolving International Legal Regime for Offshore Accidental Pollution Prevention, Preparedness, and Response' (2013) 44(4) *Ocean Development and International Law* 335–362
Vinogradov, Sergei and Patricia Wouters, 'The Caspian Sea: Quest for a New Legal Regime' (1996) 9 *Leiden Journal of International Law* 94
Vinokurov, Evgeny and Alexander Libman, *Eurasian Integration: Challenges of Transcontinental Regionalism* (Palgrave Macmillan, Basingstoke, 2012)
Viñuales JE and E Lees (eds), *Environmental and Energy Law* (Edward Elgar, Cheltenham, in press)
Vuik, Kees, 'Is it Finally Possible to Predict Earthquakes?', *The Guardian*, 20 May 2015
Waibel, Gabi, Simon Benedikter, Nadine Reis, Sven Genschick, Loan Nguyen, Pham Cong Huu and Tran Tranh Be, 'Water Governance under Renovation? Concepts and Practices of IWRM in the Mekong Delta, Vietnam', in FG Renaud and C Kuenzer, *The Mekong Delta System: Interdisciplinary Analyses of a River Delta* (Springer, 2012) 167–198
Walker, Beth, 'China Gives Green Light to New Era of Mega-Dams', in China Dialogue, *Forging a New Course for the Mekong* 4–5
Wan, William, 'Study: Pollution from Chinese Factories is Harming Air Quality on US West Coast', *The Washington Post*, 21 January 2014
Wang Canfa, 'Chinese Environmental Law Enforcement: Current Deficiencies and Suggested Reforms' (2006–2007) 8 *Vermont Journal of Environmental Law* 163
Wang, Shi-jun, Wang Dan and Yang Xiang-hua, 'Urbanization and its Impacts on Water Environment in Tumen River Basin' (2002) 12 *Chinese Geographical Science* 273–281

Warner, Robin, 'Stemming the Black Tide: Cooperation on Oil Pollution Preparedness and Response in the South China Sea and East Asian Seas' (2015) 18(2) *Journal of International Wildlife Law* 184–197

Warner, Robin, 'Conserving Marine Biodiversity in the Global Marine Commons: Co-Evolution and Interaction with the Law of the Sea' (2014) 6(1) *Frontiers in Marine Science* 1–23

Warner, Robin and Simon Marsden (eds), *Transboundary Environmental Governance: Inland, Coastal and Marine Perspectives* (Ashgate, Farnham, UK, 2012)

Warner, Robin and Simon Marsden, 'Perspectives on Transboundary Environmental Governance', in Robin Warner and Simon Marsden (eds), *Transboundary Environmental Governance: Inland, Coastal and Marine Perspectives* (Ashgate, Farnham, UK, 2012) 1–17

Waslekar, Sundeep, 'Asia's Water Can be a Source of Harmony', *South China Morning Post*, 11 January 2016

Wearden, Graeme, 'Iran Urges Other Oil Producers to Freeze Output after Doha Debacle – as it Happened', *The Guardian*, 19 April 2016

Webb, Simon, 'ASEAN Says Seriously Concerned About Rising South China Sea Tensions', *Reuters*, 27 February 2016

Wegerich, Kai, 'Water Resources in Central Asia: Regional Stability or Patchy Make-Up?' (2011) 30(2) *Central Asian Survey* 275–290

Weinthal, Erika, *State Making and Environmental Cooperation: Linking Domestic and International Politics in Central Asia* (MIT Press, Cambridge, 2002)

Wishart, Ruth, 'The EU Referendum is Already Following the Scottish Playbook, Project Fear 2.0', *The Guardian* 26 February 2016

Wolf, Aaron T and Joshua T Newton, 'Case Study of Transboundary Dispute Resolution: The Indus Waters Treaty', undated

Won Bae Kim, Yue-man Yeung and Sang-chuel Choe, *Collaborative Regional Development in NE Asia: Towards a Sustainable Regional and Sub-regional Future* (Chinese University of Hong Kong Press, 2011)

Wong, Catherine, 'Water Releases "Won't Wash Away Mistrust" Over Mekong', *South China Morning Post*, 18 April 2016

Wong, Edward, 'China Region Bans Glacier Tourism', *International New York Times*, 18 February 2016

Wong, Peter, 'The Dealmaker', *South China Morning Post*, 30 May 2016

World Bank, Operational Manual, OP 4.01 – Environmental Assessment, 2013

Wouters, Patricia, 'The Yin and Yang of International Water Law: China's Transboundary Water Practice and the Changing Contours of State Sovereignty' (2014) 23(1) *Review of European Community and International Environmental Law* 67–75

Wouters, Patricia and H Chen, 'China's "Soft Path" to Transboundary Water Cooperation Examined in the Light of Two UN Global Water Conventions – Exploring the "Chinese Way"' (2014) (22) *The Journal of Water Law* 229–247

Wouters, Patricia and Kenneth Kang, 'Water Rights and Social Justice in the Mekong Region' (2013) 33(1) *Mountain Research and Development* 116–117

Xiuzhen, Che, Shang, Jincheng and Wang, Jinhu, 'Strategic Environmental Assessment and its Development in China' (2002) 22 *Environmental Impact Assessment Review* 101–109

Xue, Hanquin, *Chinese Contemporary Perspectives on International Law: History, Culture and International Law* (Brill Nijhoff, Leiden, 2012)

Yang, Jiang, 'Silk Road Feels the Heat', *South China Morning Post*, 8 December 2014

Yang, Shanshan, 'Public Participation in the Chinese Environmental Impact Assessment (EIA) System' (2008) 10 *Journal of Environmental Assessment Policy and Management* 91–113

Yang, SZ, *Central Asia in Transformation and China* (Beijing, 2005)

Ye, Amy Aai Sheau, 'Liability and Compensation Regime for Transboundary Oil Pollution Damage' (2014) 5(2) *Australian Journal of Maritime and Ocean Affairs* 59–64

Yeh, Emily T, *Taming Tibet: Landscape Transformation and the Gift of Chinese Development* (Cornell University Press, Ithaca and London, 2013)

Yeung, SC, 'Is HK Independence a Mainstream Thought among the Youth?', *Economic Journal Insight*, 24 February 2016

Yoon, Esook, 'Governing Transboundary Pollution in Northeast Asia', in Jörg Balsiger and Aysun Uyar (eds), *Comparing Regional Environmental Governance in East Asia and Europe: Proceedings* (Research Institute for Humanity and Nature, Kyoto, 2013) 65–70

Yoshimatsu, Hidetaka, 'Regional Governance and Cooperation in Northeast Asia: The Cases of Environment and IT', RCAPS Working Paper No. 09-9, Ritsumeikan Center for Asia Pacific Studies (RCAPS), Ritsumeikan Asia Pacific University, March 2010

Young, Oran R (ed), *The Effectiveness of International Environmental Regimes* (MIT Press, Cambridge, 1999)

Young, Oran R, *International Cooperation: Building Regimes for Natural Resources and the Environment* (Cornell University Press, Ithaca and London, 1989)

Zawahri, Neda A and Oliver Hensengerth, 'Domestic Environmental Activists and the Governance of the Ganges and Mekong Rivers in India and China' (2012) 12 *International Environmental Agreements* 269–298

Zhu Tan and Kin-Che Lam (eds), *Environmental Impact Assessment in China* (Nankai University and Chinese University of Hong Kong, 2009)

Zhuang Pinghui, 'Key Facts Behind China's Warming Ties with Saudi Arabia, Iran and Egypt as Xi Jinping Signs Mega Oil Deals During his Middle East Tour', *South China Morning Post*, 20 January 2016

Ziganshina, Dinara, 'International Water Law in Central Asia: Commitments, Compliance and Beyond' (2009) *The Journal of Water Law* 96–107

Zonn, I, A Kostianoy, A Kosarev and M Glanz, *The Caspian Sea Encyclopedia* (Springer Verlag, Berlin, 2011)

Zuo, Mandy, 'The Road from China's Past to Future', *South China Morning Post*, 8 December 2014

Index